D1426634

UNIVERSITY OF NOTTINGHAM
TELEPEN
6 30 011139 5

INTERNATIONAL SERIES OF MONOGRAPHS ON
CHEMICAL ENGINEERING
GENERAL EDITORS: P. V. DANCKWERTS and R. F. BADDOUR

VOLUME 1

NON-NEWTONIAN FLUIDS

FLUID MECHANICS, MIXING AND HEAT TRANSFER

NON-NEWTONIAN FLUIDS

FLUID MECHANICS, MIXING AND HEAT TRANSFER

by

W. L. WILKINSON, M.A., Ph.D., A.M.I. Chem. E.

Lecturer in Chemical Engineering at University College, Swansea

PERGAMON PRESS

NEW YORK · LONDON · OXFORD · PARIS

1960

PERGAMON PRESS INC.
122 East 55th Street, New York 22, N.Y.
1404 New York Avenue, N.W., Washington 5, D.C.
P.O. Box 47715, Los Angeles, California

PERGAMON PRESS, LTD.
4 & 5 Fitzroy Square, London, W.1
Headington Hill Hall, Oxford

PERGAMON PRESS, S.A.R.L.
24 Rue des Écoles, Paris Ve

PERGAMON PRESS, G.m.b.H.
Kaiserstrasse 75, Frankfurt am Main

Copyright
©
1960
PERGAMON PRESS LTD.

Library of Congress Card No. 59–14174

Printed in Great Britain by Page Bros. (Norwich) Ltd.

CONTENTS

PREFACE

In fluid mechanics the simple Newtonian fluid has come to be regarded as normal and fluids which show deviations from this type of flow behaviour have been considered anomolous. Unfortunately, these so-called anomolous or *non-Newtonian* fluids are all too frequently encountered in the chemical and process industries and it is becoming increasingly important for the engineer to be familiar with the special problems involved. In fact, one could easily visualize a much broader approach to fluid mechanics into which the Newtonian fluid fits as a comparatively inconspicuous special case.

There are several quite different approaches to a study of non-Newtonian systems. The problem of the physical chemist is the interpretation of the flow behaviour of a fluid in terms of its physical and chemical properties. The theoretical rheologist sets up a more or less complicated mathematical model of a fluid and proceeds to derive its behaviour from this basis. Most of the early work has been of one or other of these types and as a result is of limited engineering significance. The present work is concerned with what might be termed the *engineering approach*, that is to say, the development of quantitative design procedures based on the measured properties of real fluids. This approach to rheology is comparatively new and so far only the more simple types of non-Newtonian fluids are amenable to this sort of treatment.

In the first two chapters the various types of non-Newtonian fluid are considered, with typical examples, and the principles of the experimental techniques normally used to characterize these materials are discussed and compared. The details of the experimental methods and suitable apparatus are to be found in the last chapter.

The remainder of the book is concerned with the fundamental principles underlying the design and operation of the equipment in which non-Newtonian materials are to be processed and the author would here like to pay tribute to the extensive pioneering work of Professor A. B. Metzner and his colleagues at the University of Delaware in this important though hitherto much neglected field.

The flow of non-Newtonian materials in pipes and channels, both in the laminar and turbulent regions, is dealt with in the third chapter. Extrusion and rolling processes have been included. Chapter 4 is concerned with the heat transfer characteristics of non-Newtonian materials. A semi-theoretical treatment has been followed for simplified situations involving heat transfer in laminar flow and then empirical formulae which are useful for rough predictions of heat transfer coefficients in turbulent non-Newtonian systems are presented are discussed. Agitation of non-Newtonian fluids is considered in Chapter 5 and here the severe lack of quantitative information is emphasised.

These engineering design procedures are restricted to the relatively simple type of non-Newtonian fluid in which the complications due to time effects and viscoelasticity are not present. Even so, this analysis is also applicable to the more complicated thixotropic and rheopectic fluids after they have been sheared for some time and reach a steady state. The same applies to visco-elastic fluids in steady laminar flow in a uniform pipe. However, when unsteady conditions exist, such as at abrupt changes of cross section and in the whole field of turbulent flow, the viscoelastic properties of these materials will be of paramount importance. At the present time virtually no work has been reported in this field in spite of the fact that many fluids of great industrial importance, such as polymers and polymer solutions, fall into this category.

The author wishes to acknowledge the assistance given to him by Mr. J. Harris in reading through the manuscript and helping in correcting the proofs.

Swansea
June 1959

W. L. WILKINSON

NOTATION

A	Area ; a constant
B	A constant ; coefficient of thixotropic breakdown
C	A constant
D	Diffusion coefficient
$\mathscr{D}$	Differential operator, d/dt ; diameter
F	Force ; function
G	Shear modulus ; couple per unit height
G^*	Dynamic shear modulus
G'	Real part of dynamic shear modulus
G''	Imaginary part of dynamic shear modulus
J_n	Bessel function of order n
J	Shear compliance, reciprocal of rigidity modulus
J^*	Dynamic shear compliance
J'	Real part of dynamic shear compliance
J''	Imaginary part of dynamic shear compliance
K	Constant
L	Length
L_e	Entrance length
M	Coefficient of thixotropic breakdown ; momentum
N	Rotational speed
P	Pressure ; exponent in Eqn. (3.7.8) ; power
Q	Volume rate of flow
R	Radius
S	Slope of logarithmic plot of torque v. speed ; reciprocal of n; distance defined in Fig. 54.
T	Dimensionless shear stress, Eqn. (3.6.4) ; temperature
U	Velocity at centre-line of a pipe ; peripheral speed of roll
V	Peripheral speed of flight of extruder screw
Z	Function
a	Radius of a pipe ; constant
b	Constant ; width of extruder channel
c	Ratio of yield stress to shear stress ; a constant
c_f	Friction factor
c_p	Specific heat at constant pressure
e	Exponential constant
f	Function
g	Defined by Eqn. (3.8.7) ; gravitational acceleration
h	Half the height of extruder flight ; heat transfer coefficient

k	Constant in power law, Eqn. (1.2.3) ; thermal conductivity
k'	Defined by Eqn. (2.4.5)
m	Defined by $m = k' \, 8^{n'-1}$
n	Exponent in power law, Eqn. (1.2.3)
n'	Exponent defined from Eqn. (2.4.2)
p	Normal stress ; Plasticity number, $\tau_y D/\mu_p u_m$
q	Heat flux
r	Radial co-ordinate
r_p	Radius of solid plug in Bingham plastic flow in a pipe
r_y	Radius of solid cylinder in Bingham plastic flow in coaxial cylinder viscometer
s	Velocity of slip
t	Time
u	Velocity ; dimensionless group Q/bhV
u_m	Mean velocity
u^+	Ratio of velocity to friction velocity, i.e. $u/\sqrt{(\tau_\omega/\rho)}$
u_p	Velocity of plug in Bingham plastic flow in a pipe
w	Mass flow rate
x	Distance ; exponent
y	Distance ; exponent
y^+	Defined by $y^+ = y\sqrt{(\tau_\omega \rho)}/\mu$
z	Distance ; dimensionless group defined by Eqn. [3.7.7]
α	Thermal diffusivity
α'	Velocity gradient at the wall of a pipe
β	Roots of $J_0(\beta a) = 0$; variable of integration
γ	Strain ; function defined in Section 3.6 (c)
$\dot{\gamma}$	$d\gamma/dt$ or rate of shear
δ	Element of length ; boundary layer thickness ; ratio of velocity gradient at wall of a pipe for non-Newtonian fluid to velocity gradient for Newtonian fluid at same flow rate ; dimensionless ratio defined by Eqn. [3.8.3]
ϵ	Roughness size ; limit of anomalous flow near a wall
ζ	Slip coefficient
η	Dimensionless group defined by Eqn. [3.8.3]
θ	Dimensionless temperature
κ	Ratio of radii of an annulus
λ	Relaxation or retardation time ; ratio
λ_i	Roots of $J_0(\lambda) = 0$
μ	Newtonian viscosity
μ_p	Bingham plastic viscosity
μ_a	Apparent viscosity, ratio of shear stress to rate of shear
μ_0	Viscosity at zero shear rate

μ_∞	Viscosity in infinite shear rate
μ_t	Turbulent viscosity
υ	Kinematic viscosity
ξ	Pressure gradient ; dimensionless group defined by Eqn. [3.8.3]
ρ	Density ; dimensionless radial distance, r/a ; coefficient defined in section 3.8
σ	Surface tension
τ	Shear stress
τ_y	Yield stress
τ_ω	Shear stress at the wall
ϕ	Creep function defined by Eqn. [2.7.1] ; dimensionless velocity defined by Eqn. [3.6.4]
ψ	Cone angle ; stress relaxation function defined by Eqn. [2.7.2]
ω	Frequency ; dimensionless group in Eqn. [3.7.7]
Δ	Incremental quantity
Ω	Angular velocity
Ω_B	Function defined by Eqn. [3.6.11]
Ω_p	Function defined by Eqn. [3.6.18]

DIMENSIONLESS RATIOS

Fr	Froude number $\frac{u^2}{gL}$
Gr	Graetz number $\frac{wC_p}{kL}$
He	Hedström number $\frac{\tau_y D^2 \rho}{\mu_p^2}$
Nu	Nusselt number $\frac{hD}{k}$
Pr	Prandtl number $\frac{\mu C_p}{k}$
Re	Reynolds number $\frac{\rho u D}{\mu}$
Re′	Generalized Reynolds number $\frac{D^{n'} u_m^{2-n'} \rho}{k' \, 8^{n'-1}}$
	Reynolds number for power law fluid $\frac{D^n u_m^{2-n} \rho}{\frac{k}{8}\left(\frac{6n+2}{n}\right)^n}$

Sc Schmidt number $\frac{\mu}{\rho \mathscr{D}}$

St Stanton number $\frac{h}{\rho u C_p}$

We Weber number $\frac{\rho u^2 L}{\sigma}$

CHAPTER 1

CLASSIFICATION OF NON-NEWTONIAN FLUIDS

1.1 GENERAL CONSIDERATIONS AND DEFINITIONS

(a) Viscosity of Newtonian Fluids

Consider a thin layer of fluid between two parallel planes a distance $\mathrm{d}y$ apart as in Fig. 1.

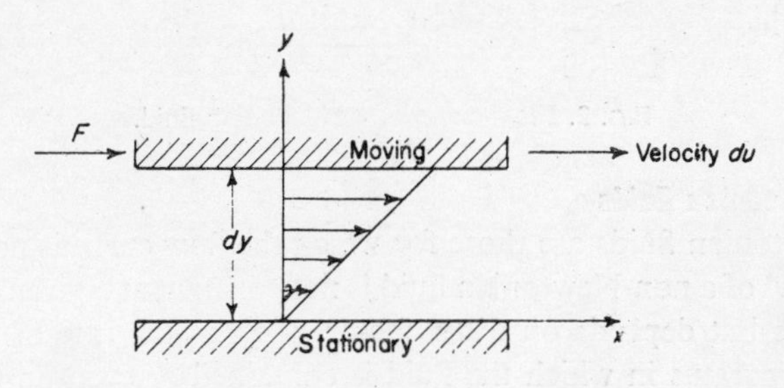

FIG. 1.

One plate is fixed and a shearing force F is applied to the other. When conditions are steady the force F will be balanced by an internal force in the fluid due to its viscosity. For a Newtonian fluid in laminar flow the shear stress is proportional to the velocity gradient, i.e.

$$F/A = \tau \propto \mathrm{d}u/\mathrm{d}y$$

This equation may be written as

$$\tau = \mu \mathrm{d}u/\mathrm{d}y = \mu \, \dot{\gamma} \qquad [1.1.1]$$

where the constant of proportionality, μ, is called the Newtonian viscosity. It will be seen that μ is the tangential force per unit area exerted on layers of fluid a unit distance apart and having a unit velocity difference between them.

The Newtonian viscosity, μ, depends only on temperature and pressure and is independent of the rate of shear. The diagram relating shear stress and rate of shear for Newtonian fluids, the so-called '*flow curve*', is therefore a straight line of slope μ as in Fig. 2, and the single constant, μ, completely characterizes the fluid.

Newtonian behaviour is exhibited by fluids in which the dissipation of viscous energy is due to the collision of comparatively small molecular species. All gases and liquids and solutions of low molecular weight come into this category. Notable exceptions are colloidal suspensions and polymeric solutions where the molecular species are large. These fluids show marked deviations from Newtonian behaviour.

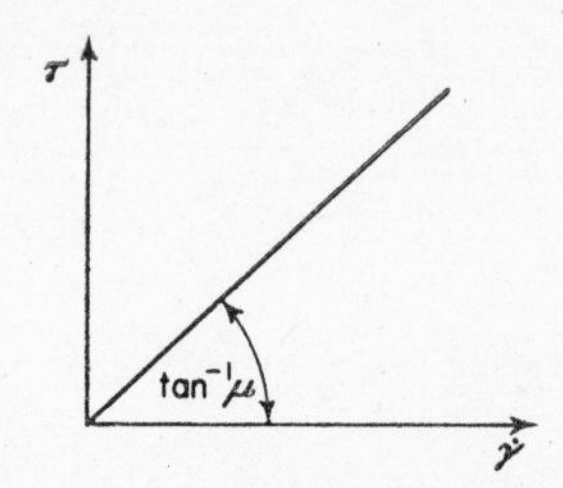

FIG. 2. Flow curve of a Newtonian fluid.

(b) Non-Newtonian fluids

Non-Newtonian fluids are those for which the flow curve is not linear, i.e. the 'viscosity' of a non-Newtonian fluid is not constant at a given temperature and pressure but depends on other factors such as the rate of shear in the fluid, the apparatus in which the fluid is contained or even on the previous history of the fluid.

These real fluids for which the flow curve is not linear may be classified into three broad types:

(1) fluids for which the rate of shear at any point is some function of the shearing stress at that point and depends on nothing else;

(2) more complex systems for which the relation between shear stress and shear rate depends on the time the fluid has been sheared or on its previous history;

(3) systems which have characteristics or both solids and fluids and exhibit partial elastic recovery after deformation, the so-called viscoelastic fluids.

These three classes of fluids will now be treated in order.

1.2 TIME-INDEPENDENT NON-NEWTONIAN FLUIDS

Fluids of the first type whose properties are independent of time may be described by a rheological equation of the form

$$\dot{\gamma} = f(\tau) \qquad [1.2.1]$$

This equation implies that the rate of shear at any point in the fluid is a simple function of the shear stress at that point. Such fluids may be termed *non-Newtonian viscous fluids.*

These fluids may conveniently be subdivided into three distinct types depending on the nature of the function in Eqn. [1.2.1]. These types are

(1) Bingham plastics
(2) pseudoplastic fluids
(3) dilatant fluids

and typical flow curves for these three fluids are shown in Fig. 3 and compared with the linear relation typical of Newtonian fluids.

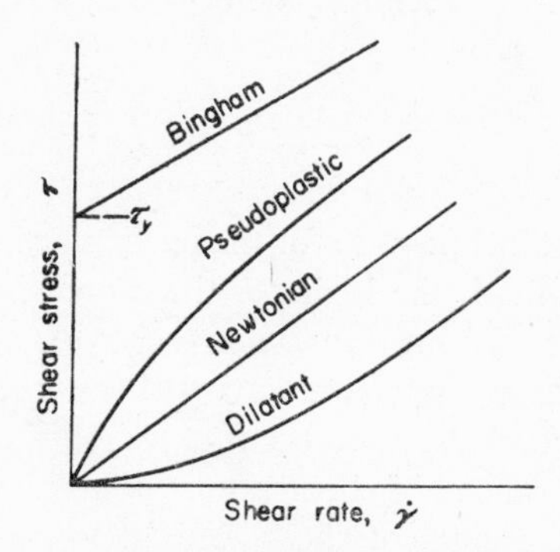

FIG. 3. Flow curves for various types of time-independent non-Newtonian fluids.

(a) Bingham plastics

A Bingham plastic[1] is characterized by a flow curve which is a straight line having an intercept τ_y on the shear-stress axis. The yield stress, τ_y, is the stress which must be exceeded before flow starts. The rheological equation for a Bingham plastic may be written

$$\tau - \tau_y = \mu_p \dot{\gamma} \; ; \; \tau > \tau_y \qquad [1.2.2]$$

where μ_p, the plastic viscosity or coefficient of rigidity, is the slope of the flow curve.

The concept of an idealized Bingham plastic is very convenient in practice because many real fluids closely approximate this type of behaviour. Common examples are slurries, drilling muds, oil paints, toothpaste and sewage sludges. The explanation of Bingham plastic behaviour is that the fluid at rest contains a three-dimensional structure of sufficient rigidity to resist any stress less than the yield stess, τ_y. If this stress is exceeded the structure completely disintegrates and the system behaves as a Newtonian fluid under a shear stress $\tau - \tau_y$. When the shear stress falls below τ_y the structure is reformed.

(b) Pseudoplastic fluids

Pseudoplastic fluids show no yield value and the typical flow curve for these materials indicates that the ratio of shear stress to the rate of shear, which may be termed the apparent viscosity, μ_a, falls progressively with shear rate and the flow curve becomes linear only at very high rates of shear. This limiting slope is known as the viscosity at infinite shear and is designated μ_∞.

The logarithmic plot of shear stress and rate of shear for these materials is often found to be linear with a slope between zero and unity. As a result, an empirical functional relation known as the *power law* is widely used to characterize fluids of this type. This relation, which was originally proposed by Ostwald [2] and has since been fully described by Reiner, [3] may be written as

$$\tau = k\,\dot{\gamma}^n \qquad [1.2.3]$$

where k and n are constants ($n < 1$) for the particular fluid: k is a measure of the consistency of the fluid, the higher k the more viscous the fluid; n is a measure of the degree of non-Newtonian behaviour, and the greater the departure from unity the more pronounced are the non-Newtonian properties of the fluid. It is important to remember that although n is nearly constant in many cases over wide ranges of shear rate it is not a true constant for real fluids over all possible ranges of shear. This is not a serious drawback in engineering applications because all that is needed is a rheological equation which describes the fluid over the particular range of shear rate encountered in the particular problem. Over such a range n may often be regarded as constant.

It should be noted here that the dimensions of k depend on the index n and this fact has led to many objections to the use of the power law, e.g. by Reiner [3]. In most engineering applications these objections are not serious.

The apparent viscosity, μ_a, for a power law fluid may be expressed in terms of n since

$$\mu_a = \tau/\dot{\gamma}$$

$$\text{i.e. } \mu_a = k\dot{\gamma}^{n-1} \qquad [1.2.4]$$

and since $n < 1$ for pseudoplastics the apparent viscosity decreases as the rate of shear increases. This type of behaviour is characteristic of suspensions of asymmetric particles or solutions of high polymers such as cellulose derivatives. This suggests that the physical interpretation of this phenomenon is probably that with increasing rates of shear the asymmetric particles or molecules are progressively aligned. Instead of the random intermingled state which exists when the fluid is at rest the major axes are brought into line with the direction of flow. The apparent viscosity continues to decrease with increasing rate of shear until no further alignment along the streamlines is possible and the flow curve then becomes linear.

Pseudoplastic fluids have been defined as time-independent fluids and this implies that the alignment of molecules suggested above takes place instantaneously as the rate of shear is increased or, at any rate, so quickly that the time effect cannot be detected using ordinary viscometric techniques.

Other empirical equations which have been used to describe pseudoplastic behaviour are

Prandtl $\tau = A \sin^{-1}(\dot{\gamma}/C)$

Eyring $\tau = \dot{\gamma}/B + C \sin(\tau/A)$

Powell-Eyring $\tau = A\,\dot{\gamma} + B \sinh^{-1}(C\dot{\gamma})$

Williamson $\tau = A\dot{\gamma}/(B + \dot{\gamma}) + \mu_{\infty}\dot{\gamma}$ [1.2.5]

In these equations A, B and C are constants which are typical of the particular fluid. These equations are considerably more difficult to use than the power law and usually do not offer any compensating advantages.

(c) Dilatant fluids

Dilatant fluids are similar to pseudoplastics in that they show no yield stress but the apparent viscosity for these materials *increases* with increasing rates of shear. The power law equation is again often applicable but in this case the index n is greater than unity.

This type of behaviour was originally found in suspensions of solids at high solids content by Osborne Reynolds. He suggested that when these concentrated suspensions are at rest the voidage is at a minimum and the liquid is only sufficient to fill these voids. When these materials are sheared at *low* rates the liquid lubricates the motion of one particle past another and the stresses are consequently small. At higher rates of shear the dense packing of the particles is broken up and the material expands or 'dilates' slightly and the voidage increases. There is now insufficient liquid in the new structure to lubricate the flow of the particles past each other and the applied stresses have to be much greater. The formation of this structure causes the apparent viscosity to increase rapidly with increasing rates of shear.

The term 'dilatant' has since come to be used for all fluids which exhibit the property of increasing apparent viscosity with increasing rates of shear. Many of these, such as starch pastes, are not true suspensions and do not dilate on shearing in the normal sense of the word. The above explanation therefore does not apply but nevertheless they are commonly referred to as 'dilatant' fluids.

In the process industries dilatant fluids are much less common than pseudoplastic fluids but when the power law is applicable the treatment of both types is much the same.

1.3 TIME-DEPENDENT NON-NEWTONIAN FLUIDS

Many real fluids cannot be described by a simple rheological equation such as Eqn. [1.2.1] which applies to fluids for which the relation between shear stress and shear rate is independent of time. The apparent viscosity of more complex fluids depends not only on the rate of shear but also on the time the shear has been applied. These fluids may be subdivided into two classes:

(*a*) thixotropic fluids

(*b*) rheopectic fluids

according as the shear stress decreases or increases with time when the fluid is sheared at a constant rate.

(a) Thixotropic fluids—breakdown of structure by shear

Thixotropic materials are those whose consistency depends on the duration of shear as well as on the rate of shear.

If a thixotropic material is sheared at a constant rate after a period of rest, the structure will be progressively broken down and the apparent viscosity will decrease with time. The rate of breakdown of structure during shearing at a given rate will depend on the number of linkages available for breaking and must therefore decrease with time. (This could be compared with the rate of a first-order chemical reaction.) The simultaneous rate of reformation of structure will increase with time as the number of possible new structural linkages increases. Eventually a state of dynamic equilibrium is reached when the rate of build-up of structure equals the rate of breakdown. This equilibrium position depends on the rate of shear and moves towards greater breakdown at increasing rates of shear.

As an example we could consider the material confined in a cylindrical viscometer (see Chapter 6). After the material has been resting for a long time one of the cylinders is rotated at a constant speed. The torque on the other cylinder would then decrease with time as shown in Fig. 4. The rate of decrease and the final torque would both depend on the speed, i.e. on the rate of shear.

Thixotropy is a reversible process and after resting the structure of the material builds up again gradually. If the flow curve of a thixotropic material is determined immediately after shearing and after it has rested for varying times after shearing the result is as in Fig. 5.

This type of behaviour leads to a kind of hysteresis loop on the curve of shear-stress plotted against rate of shear if the curve is plotted first for the rate of shear increasing at a constant rate and then for the rate of shear decreasing at a constant rate. This is illustrated in Fig. 6, where the curves A and B are drawn for fluids of the Newtonian and pseudoplastic types which exhibit thixotropy.

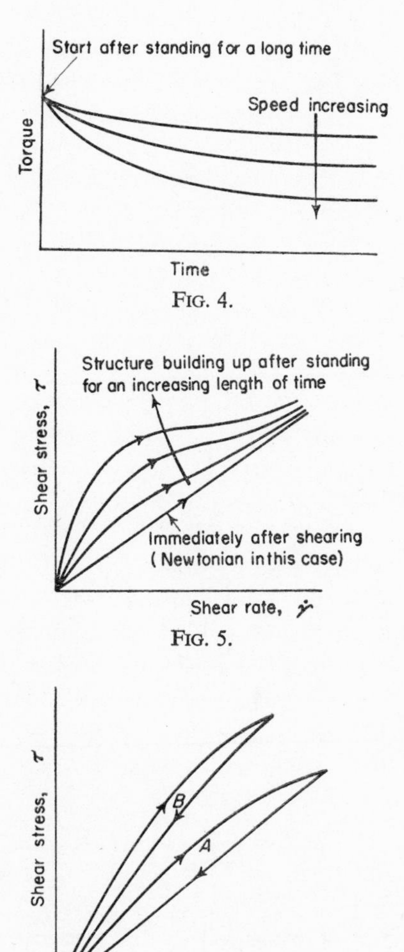

FIG. 4.

FIG. 5.

FIG. 6. Hysteresis loops for thixotropic fluids.

Flow curves of increasing height can be obtained by applying shear for increasing lengths of time before making the return path. A single curve can be obtained by continuing the shearing process to equilibrium before returning.

The term 'false-body' is frequently encountered in discussions on thixotropy. This was introduced by Pryce-Jones [4] to distinguish types of thixotropic behaviour of Bingham plastics. True thixotropic materials break down completely under the influence of high stresses and behave like true liquids even after the stress has been removed, until such time as the structure has reformed. False-bodied materials, on the other hand, do not lose their solid properties entirely and can still exhibit a yield value even though this might be diminished. The original yield value is only regained after resting for a long time.

The hysteresis loop on the flow curve would take the form of Fig. 7 for these two materials.

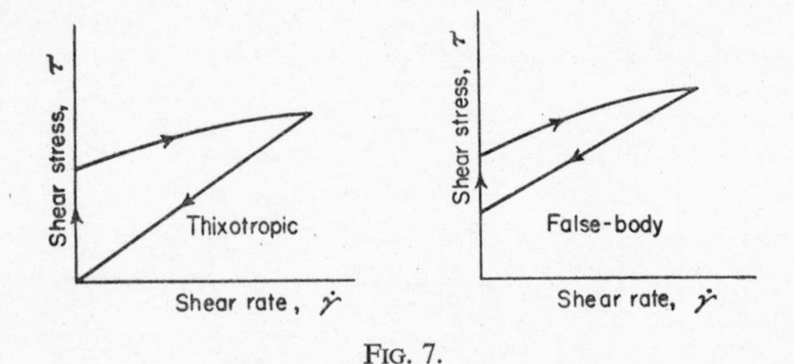

FIG. 7.

This behaviour can be illustrated by the following experiment. Consider a liquid in a vessel with a cylinder on a torsion wire immersed in it. The cylinder is deflected and the liquid stirred. Stirring is then stopped and the cylinder released. The torsion in the wire would then vary with time as in Fig. 8 for the two types of material.

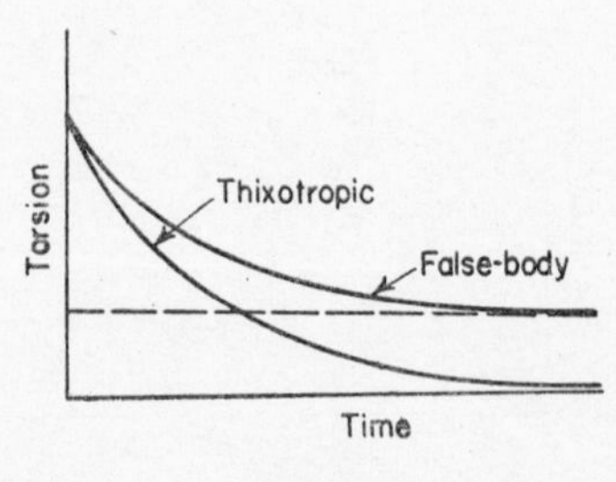

FIG. 8.

With the false-bodied materials there would be a residual torsion in the wire indicating that the material can offer permanent resistance to shear immediately after stirring, i.e. it retains a finite yield value. A true thixotropic material would show a residual torsion only if some time elapsed after stirring had ceased before the cylinder was released, to allow structure to build-up.

In a like fashion, false-bodied materials can retain elasticity (see Section 1.4). This would result in a recoil of the cylinder in the above experiment.

(b) Rheopectic fluids—formation of structure by shear

This is a case of gradual formation of structure by shear, whereas so far the properties of structured materials have been explained on the basis that shearing tends to destroy structure.

Freundlich and Juliusberger,[5] using a 42 per cent gypsum paste ($1 - 10\mu$) in water, found that after shaking, this material re-solidified in 40 min if at rest, but in 20 sec if the vessel was gently rolled in the palms of the hands. This seems to indicate that small shearing motions facilitate structure build-up but large shearing (shaking) destroys it. There is a critical amount of shear beyond which reformation of structure is not induced but breakdown occurs. This behaviour is also observed with dilute aqueous solutions of vanadium pentoxide and bentonite.

There are other materials, however, in which structure only forms under shear and gradually disintegrates when at rest. This is usually termed 'rheopexy' but it is quite distinct from the definition of this term given by Freundlich to the behaviour of gypsum pastes. Even so this behaviour is only found at moderate rates of shear, for if shearing is rapid the structure does not form. A 0·005 N suspension of ammonium oleate behaves in this way. Consider the flow of this material through a capillary tube. At a moderate pressure difference the flow is rapid at first and then decreases as the structure builds up. At a high pressure difference the flow is always rapid and does not fall off because the structure does not build up at high rates of shear.

(c) Relation between time-dependent and time-independent fluids

Thixotropy is rather like pseudoplasticity in which the time required for the alignment of particles is not negligible. This time effect for 'pseudoplastic materials' is not observable in the apparatus normally used for the testing of these fluids. The difference then is only a matter of degree.

In the same way rheopectic fluids (e.g. ammonium oleate) are superficially similar to their time-independent counterparts (dilatant fluids) in which the time for structure build-up is insignificantly small. Here, however, the analogy is not so close because rheopexy is a case where build-up is brought about by small shearing rates only. There is an upper limit to the shear rate beyond which the analogy breaks down.

1.4 VISCOELASTIC FLUIDS

A viscoelastic material is one which possesses both elastic and viscous properties, i.e. although the material might be viscous, it exhibits a certain elasticity of shape. This concept is perhaps most easily visualized in the case of a very viscous liquid such as pitch. Suppose initially we consider the simplest case where we assume Newton's law for the viscous component and Hooke's law for the elastic component. In steady state flow under a shear stress τ the rate of shear will be τ/μ_0 where μ_0 is a constant Newtonian viscosity coefficient. Suppose now that the shear stress is increased to $\tau + \delta\tau$ very rapidly. The material will now be sheared through an additional angle $\delta\tau/G$ where G is a rigidity modulus. There is therefore an additional rate of shear proportional to the rate of change of stress at any instant and the total rate of shear is given by

$$\dot{\gamma} = \tau/\mu_0 + \dot{\tau}/G \qquad [1.4.1]$$

or we can write this as

$$\tau + \lambda_1 \dot{\tau} = \mu_0 \dot{\gamma} \text{ where } \lambda_1 = \mu_0/G \qquad [1.4.2]$$

This equation was first proposed by Maxwell,[6] and liquids which are described by it are usually referred to as 'Maxwell liquids'.

The parameter λ_1 has dimensions of time and it is seen from Eqn. [1.4.2] that it is the time constant of the exponential decay of stress at a constant strain, i.e. if the motion is stopped the stress will relax as $\exp(-t/\lambda_1)$. Consequently λ_1 is known as the relaxation time. Schofield and Scott-Blair[7] have successfully applied the Maxwell equation to flour doughs.

Oldroyd[8] investigated the elastic and viscous properties of emulsions and suspensions of one Newtonian liquid in another and derived theoretically the differential equation relating the shear stress τ and the rate of shear $\dot{\gamma}$ in the form

$$\tau + \lambda_1 \dot{\tau} = \mu_0(\dot{\gamma} + \lambda_2\ddot{\gamma}) \qquad [1.4.3]$$

where the constants μ_0, λ_1, and λ_2 can be determined in terms of the physical properties of the mixture. In this system the elastic strain energy is stored during flow by virtue of the fact that interfacial tension provides the restoring force which makes the individual drops resist changes of shape. The same equation was also derived by Frölich and Sack[9] for a dilute suspension of solid particles in a viscous liquid. Elastic strain energy is stored because the solid elastic particles are deformed by the flow of the surrounding liquid.

In this equation the constant μ_0 can be identified as the viscosity at low rates of shear in the steady state, i.e. when $\dot{\tau} = \ddot{\gamma} = 0$. The constant λ_1 is a relaxation time and the physical significance of this is that if the motion is suddenly stopped the shear stress will decay as $\exp(-t/\lambda_1)$; λ_2 is called a 'retardation time' and has the significance that if all stresses are removed the rate of strain decays as $\exp(-t/\lambda_2)$.

Toms and Strawbridge[10] found that the behaviour of dilute solutions of poly-methylmethacrylate in pyridine can be described by means of an equation of this sort. It also describes the behaviour of some bitumens.

It is apparent, then, that a viscoelastic fluid cannot be characterized by a simple rheological equation of the form of $\dot{\gamma} = f(\tau)$. The essential difference is that the rheological equation contains the time derivatives of both τ and γ in general. The general case may be written

$$f_1(\mathrm{D})\,\tau = f_2(\mathrm{D})\,\gamma \qquad [1.4.4]$$

or alternatively as a polynomial in D

$$\sum_{n=0}^{N} \alpha_n \mathrm{D}^n\,\tau = \sum_{m=0}^{M} \beta_m\,\mathrm{D}^m\,\gamma$$

where D is the differential operator, $\mathrm{d}/\mathrm{d}t$.

1.5 MECHANICAL ANALOGIES TO VISCOELASTIC FLUIDS

The rheological equation for a viscoelastic fluid

$$f_1(\mathrm{D})\,\tau = f_2(\mathrm{D})\,\gamma$$

is general and if solved subject to the correct boundary conditions will give the response of the material to any imposed stress or strain. However for real fluids the equations are very difficult to solve, even assuming that the values of the relevant parameters can be derived from experiment; but a great deal of qualitative information can be derived from a study of idealized mechanical models or analogies which are designed to duplicate, more or less closely, the observed time-dependence of a real fluid. Their behaviour is more easily visualized than that of a fluid, perhaps especially by engineers. A study of models also suggests a valuable method of characterizing a fluid by a single parameter. This is discussed later in this section.

These models are made up of combinations of springs and dash-pots. The force on a spring is proportional to strain and the force on a dash-pot is proportional to rate of strain. Consequently the springs and dash-pots in a model represent the elastic and viscous properties of the fluid respectively. The basic elements in any mechanical model are a parallel combination of a spring and dash-pot, known as a Voigt element, and a series combination, known as a Maxwell element (since its equation is the same as that of a Maxwell body discussed previously). These elements represent the behaviour of idealized materials. Real fluids will consist of a more or less complicated combination of these basic elements. The models for complex materials will be derived by first considering the two basic elements and then generalizing them.

(a) Voigt body

The mechanical analogy of a Hookean solid is a spring, and that of a Newtonian liquid is a dash-pot. If we combine these in parallel we have what is known as a Voigt body.

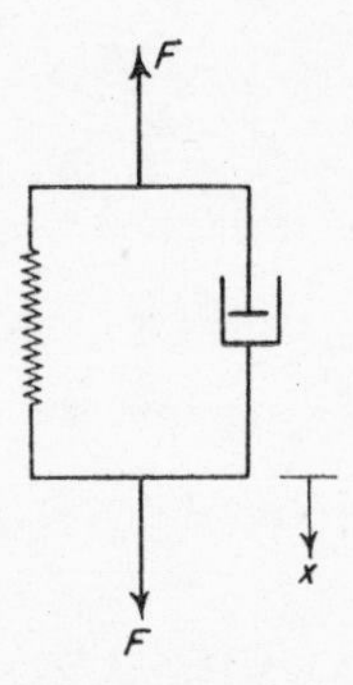

FIG. 9. Voigt body.

The equation of motion of this body is

$$F = k_1 x + k_2 \dot{x}$$

where k_1 is the spring constant and k_2 is the damping constant of the dash-pot.

If we regard the force as analogous to stress and the extension as analogous to strain we could say that this body is a mechanical analogy for a fluid whose behaviour in shear is described by the equation

$$\tau = G\gamma + \mu\,\dot{\gamma} \qquad [1.5.1]$$

where μ is a viscosity and G is a rigidity modulus.

Integrating Eqn. [1.5.1] we get in general

$$\gamma = \exp\left(-\frac{G}{\mu}t\right)\left[\gamma_0 + \frac{1}{\mu}\int \tau \exp\left(\frac{G}{\mu}t\right)\mathrm{d}t\right] \qquad [1.5.2]$$

where γ_0 is the strain at $t = 0$.

If the stress is constant at τ_0 and the initial strain is zero we have the simplified case

$$\gamma(t) = \frac{\tau_0}{G}\left[1 - \exp\left(-t/\lambda\right)\right] \qquad [1.5.3]$$

where $\lambda = \mu/G$ is the retardation time. If the stress is removed the strain vanishes exponentially with time constant λ. (This means that the strain falls to $1/e$ of its initial value in time λ.)

It should be noted here that a Voigt body is really a viscoelastic 'solid' since it can be seen that it does not exhibit unlimited non-recoverable viscous flow. It will come to rest in fact when the spring has taken up the load.

(b) Maxwell body

The Maxwell body consists of a spring and dash-pot in series. The equation of this body in rheological terms is

$$\dot{\gamma} = \dot{\tau}/G + \tau/\mu$$

and this is seen to be the same as the Maxwell body considered previously.

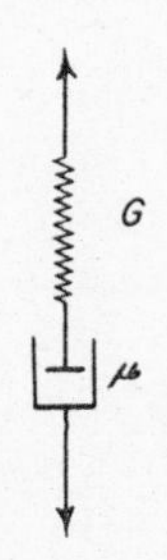

FIG. 10. Maxwell body.

Integrating we get

$$\tau = \exp\left(-\frac{G}{\mu}\,t\right)\left[\tau_0 + G\int \gamma \exp\left(\frac{G}{\mu}\,t\right)\mathrm{d}t\right] \qquad [1.5.4]$$

where τ_0 is the stress at time zero.

If a Maxwell body is subjected at $t = 0$ to a constant strain, γ_0 the stress will decay as

$$\tau(t) = \gamma_0 G \exp\left(-t/\gamma\right) \qquad [1.5.5]$$

where $\gamma = \mu/G$ is the relaxation time.

It is seen that the Maxwell body is a viscoelastic 'liquid' since there will be continuous steady flow under a given stress, τ, given by $\dot{\gamma} = \tau/\mu$ as in the case of a Newtonian liquid.

(c) Extensions of the Simple Voigt and Maxwell bodies

The simple Voigt and Maxwell bodies are not sufficiently general to describe adequately the behaviour of real viscoelastic materials. In order to extend these models into more realistic systems capable of describing real fluids it is convenient to consider a number of similar Voigt elements connected in series or a set of Maxwell bodies in parallel. (This method is used because Voigt elements in parallel exhibit the same charactristics as a single Voigt element and likewise Maxwell bodies in series behave like a single Maxwell body.)

(d) The generalized Voigt body

Consider a set of N Voigt elements in series. The nth element has a modulus G_n and a viscosity μ_n. Hence the retardation time of this element, λ_n, is given by μ_n/G_n.

Consider a stress, τ_0, suddenly applied at $t = 0$ and then kept constant. For the nth element we have from Eqn. [1.5.3]

$$\gamma_n(t) = \frac{\tau_0}{G_n}\left[1 - \exp\left(-t/\lambda_n\right)\right] \qquad [1.5.6]$$

The total strain, γ, will be the sum of the strains of all the elements; hence

$$\gamma(t) = \tau_0 \sum_{n=1}^{N} \frac{1}{G_n}\left[1 - \exp\left(-t/\lambda_n\right)\right] \qquad [1.5.7]$$

It is customary to write $1/G_n = J_n$ where J_n is called the compliance; hence

$$\gamma(t) = \tau_0 \sum_{n=1}^{N} J_n\left[1 - \exp\left(-t/\lambda_n\right)\right] \qquad [1.5.8]$$

In the limit as $N \rightarrow \infty$ there will be an infinite set of elements over which the retardation times range continuously from zero to infinity. The finite set of constants for the elements (J_n, λ_n) is now replaced by the distribution function $J(\lambda)$ which gives the amount of elastic compliance associated with the retardation time λ (now considered as a continuous parameter). This is usually referred to as a distribution of retardation times.

Eqn. [1.5.8] then becomes in the limit as $N \to \infty$

$$\gamma(t) = \tau_0 \int_0^\infty J(\lambda)\,[1 - \exp(-t/\lambda)]\,d\lambda \qquad [1.5.9]$$

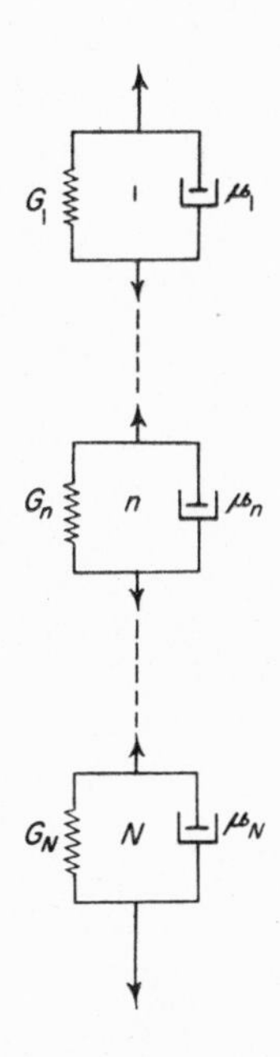

FIG. 11. Generalized Voigt body.

This concept of a distribution of retardation times greatly simplifies the mathematical approach to the problem since we can now describe the behaviour of the material by specifying its compliance distribution function $J(\lambda)$. This has been found successful in describing the behaviour of amorphous linear high polymers such as polystyrene.

It is seen that, as in the case of a simple Voigt body, the generalized Voigt model is analogous to a viscoelastic 'solid' and the system does not exhibit unlimited non-recoverable viscous flow if $G_1 \ldots G_N$ and $\mu_1 \ldots \mu_N$ are all positive. If one of the Voigt elements has a zero modulus we are left with a simple dash-pot which does allow unlimited flow, and the system could then be considered a visco-elastic liquid. Similarly if one of the dash-pots has zero viscosity the model would have some instantaneous elastic compliance.

The *creep function*, $\phi(t)$, of a viscoelastic material is defined as the strain per unit stress expressed as a function of time when the relaxed material is suddenly subjected to a constant stress. If the strain $\gamma(t)$ results after a stress τ_0 has been imposed at time zero, the creep function is given by

$$\phi(t) = \gamma(t)/\tau_0 \qquad [1.5.10]$$

For the generalized Voigt model it is seen that

$$\phi(t) = \int_0^\infty J(\lambda)[1 - \exp(-t/\lambda)d\lambda] \qquad [1.5.11]$$

and the distribution of retardation times, $J(\lambda)$, can be obtained from an experimental determination of $\phi(t)$. This is discussed later, but it should be noted here that in principle $J(\lambda)$ can be obtained from $\phi(t)$ by an inverse Laplace transformation.

(e) The generalized Maxwell model

The simple Maxwell body does not describe real viscoelastic liquids adequately in most cases, but it is possible to generalize it in much the same way as the Voigt element.

Consider the extension to a set of N elements in parallel.

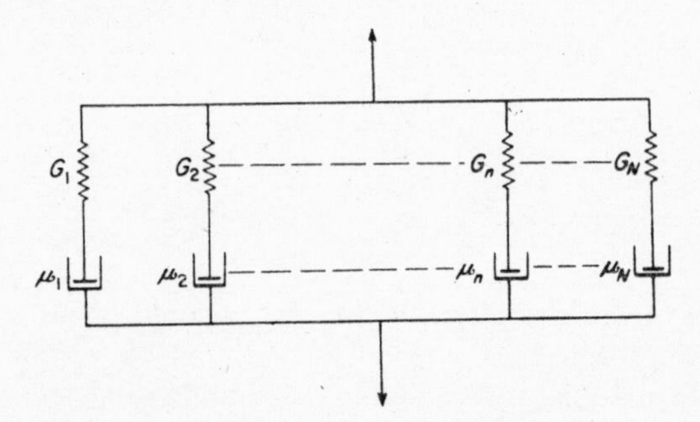

FIG. 12. Generalized Maxwell body.

For the nth element the stress is related to strain by the equation

$$\dot{\gamma}(t) = \tau_n(t)/\mu_n + \dot{\tau}_n(t)/G_n \qquad [1.5.12]$$

The total stress $\tau(t)$ will be the sum of the individual stresses, i.e,

$$\tau(t) = \sum_{n=1}^{N} \tau_n(t) \qquad [1.5.13]$$

Hence if we consider the set of N elements subjected at time zero to a deformation, γ_0, which is then kept constant, the stress would be given by

$$\tau(t) = \gamma_0 \sum_{n=1}^{N} G_n \exp(-t/\lambda_n) \qquad [1.5.14]$$

in accordance with Eqn. [1.5.5].

In the limit as $N \to \infty$ the constants (G_n, λ_n) are replaced by the distribution function $G(\lambda)$ which gives the amount of elastic modulus associated with the relaxation time, λ. This is called the 'distribution of relaxation times'.

Hence as $N \to \infty$ we have

$$\tau(t) = \gamma_0 \int_0^\infty G(\lambda) \exp(-t/\lambda)\, d\lambda \qquad [1.5.15]$$

This is the equation of the generalized Maxwell liquid.

The *relaxation function*, $\psi(t)$, of a viscoelastic material is defined as the stress per unit initial strain expressed as a function of the time t when the material has been subjected to an instantaneous strain γ_0 at time zero, i.e.

$$\psi(t) = \tau(t)/\gamma \qquad [1.5.16]$$

and for the generalized Maxwell body we have

$$\psi(t) = \int_0^\infty G(\lambda) \exp(-t/\lambda)\, d\lambda \qquad [1.5.17]$$

It is possible to derive $G(\lambda)$ from relaxation experiments in much the same way as $J(\lambda)$ can be derived from creep experiments. In principle $G(\lambda)$ can be derived from $\psi(t)$ by an inverse Laplace transformation just as $J(\lambda)$ can be found from $\phi(t)$. However, in practice these direct transformations are not feasible because the experimental relaxation or creep data are not sufficiently complete.

(f) Relationships between models and creep and relaxation functions

Since either $J(\lambda)$, $G(\lambda)$, $\phi(t)$ or $\psi(t)$ can be used to characterize a viscoelastic material it is obvious that there must be a close relationship between them. These relationships have been fully described by Alfrey.[11] They may be illustrated diagrammatically as shown in Fig. 13.

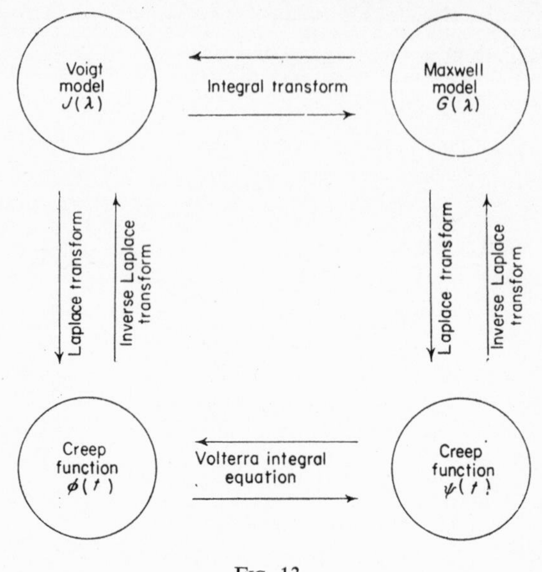

FIG. 13.

(g) Example of the use of models to describe a real fluid

It has been mentioned previously (page 10) that Oldroyd[8] has derived theoretically an equation of the form

$$\tau + \lambda_1 \dot{\tau} = \mu_0(\dot{\gamma} + \lambda_2 \ddot{\gamma})$$

for emulsions and supensions and that Strawbridge and Toms [10] have found that this equation describes the behaviour of certain polymer solutions.

Consider the following model.

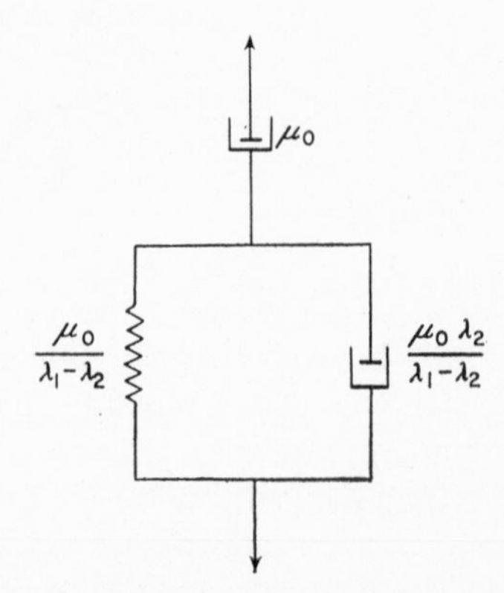

Let the strain of the Voigt element be $\gamma_1(t)$ and that of the dashpot be $\gamma_2(t)$.
For the Voigt element we have

$$\tau = G\gamma + \mu\dot{\gamma}$$

and substituting for G and μ for this model we get

$$\gamma_1(t) = \tau(t) \bigg/ \left(\frac{\mu_0}{\lambda_1 - \lambda_2} + \frac{\mu_0 \lambda_2 \mathrm{D}}{\lambda_1 - \lambda_2}\right); \; \mathrm{D} = \frac{\mathrm{d}}{\mathrm{d}t}$$

also

$$\gamma_2(t) = \tau(t)/\mu_0 \mathrm{D}$$

The total strain $\gamma(t) = \gamma_1(t) + \gamma_2(t)$, i.e.

$$\gamma(t) = \tau(t)\left[\frac{1}{\dfrac{\mu_0}{\lambda_1 - \lambda_2}(1 + \lambda_2 \mathrm{D})} + \frac{1}{\mu_0 \mathrm{D}}\right]$$

which may be written as

$$\gamma(t) = \tau(t)\,\frac{1 + \lambda_1 \mathrm{D}}{\mu_0 \mathrm{D}\,(1 + \lambda_2 \mathrm{D})}$$

or

$$\tau + \lambda_1 \dot{\tau} = \mu_0 (\dot{\gamma} + \lambda_2 \ddot{\gamma})$$

which is the same as Oldroyd's equation.

The following model can also be shown to have the same governing equation.

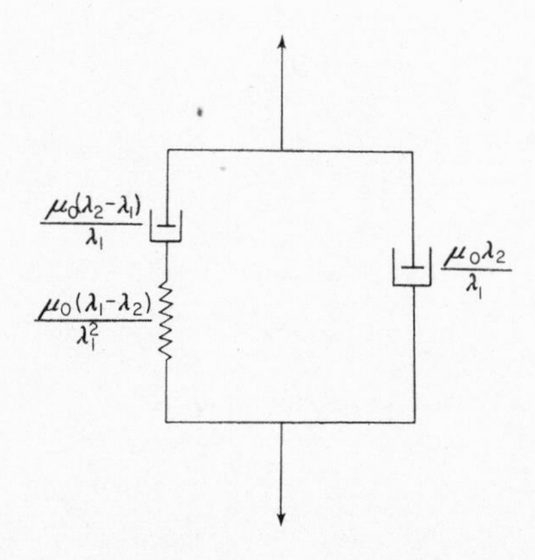

CHAPTER 2

EXPERIMENTAL CHARACTERIZATION OF NON-NEWTONIAN FLUIDS

It has been seen that at least two experimental measurements must be made on any non-Newtonian fluid in order to define its rheological properties whereas a Newtonian fluid requires only one measurement, namely the viscosity. The techniques for the experimental characterization of the three main classes of fluid, time-independent, time-dependent and visco-elastic, will be considered separately. Only the principles will be given in this chapter. Details of the apparatus required and the experimental techniques will be presented in Chapter 6.

2.1 METHODS AVAILABLE FOR THE CHARACTERIZATION OF TIME-INDEPENDENT FLUIDS

There are two main viscometric methods for the determination of the rheological properties of these fluids, as follows.

(1) Direct determination of the relation between shear stress and shear rate by subjecting the entire sample to a uniform rate of shear in a suitably designed instrument and measuring the corresponding shear stress. Viscometers using this principle are usually rotational instruments of the coaxial cylinder or cone and plate type.

(2) To infer the relation between shear stress and rate of shear indirectly from observations on the pressure gradient and volumetric flow rate in a straight pipe or capillary-tube viscometer. In these instruments the rate of shear is not constant but varies from zero at the centre of the pipe to a maximum at the wall and consequently the interpretation of the results is not so obvious.

In the following it will be assumed that there is no anomalous flow near a wall, i.e. no 'slip'. This assumption is not always valid for non-Newtonian fluids and often the rate of shear may not be a unique function of the shear stress even for time-independent fluids, because of the fact that the wall introduces a preferred orientation of those molecules or particles which are close to it and this produces in effect a slip-velocity at the walls. This topic will be discussed in more detail in Appendix 1.

2.2 TIME-INDEPENDENT FLUIDS IN ROTATIONAL INSTRUMENTS

(a) Coaxial cylinder viscometers

The principle of the coaxial-cylinder viscometer is shown in Fig. 14. The material is confined between long vertical coaxial cylinders, one of which can be rotated at various speeds while the torque on the other is measured. The variation of torque with speed can be interpreted to give the relation between shear stress and rate of shear. The variation of the rate of shear throughout the sample depends on the width of the annular gap between the two cylinders and if this is kept small the variation in shear rate can be made insignificant. The instrument is usually placed in a constant temperature bath.

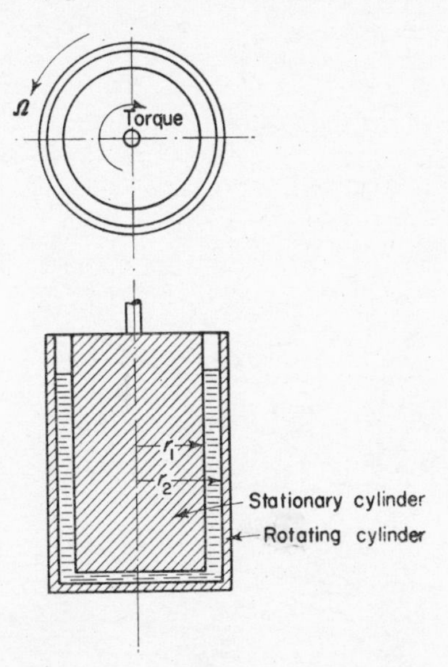

Fig. 14.

This type of instrument is subject to an end effect at the bottom of the cylinders where there is no free surface. By suitable design this can be made small but it cannot be eliminated completely. The end effect can, however, be eliminated from the result by carrying out two experiments in which the speeds of rotation are the same, but the heights of the fluid in the gap are different. If the difference between the measured torques is divided by the difference in the heights of fluid in the two experiments, the torque per unit height in the region of uniform flow is obtained. This method, though accurate, is tedious because it involves two measurements at each speed of rotation. An alternative approximate method is to calculate an 'equivalent height' corresponding to a certain height in the gap by calibrating the instrument with liquids of known viscosity.

It is shown in Appendix 2 that, for a time-independent fluid with no slip at the wall, the general functional relation between the measured torque per unit height of liquid, G, and the angular velocity of the outer cylinder, Ω, is given by

$$\Omega = \int_{r_1}^{r_2} f\left(\frac{G}{2\pi r^2}\right)\frac{\mathrm{d}r}{r} \qquad [2.2.1]$$

where the function f is that relating the shear stress τ to the rate of shear $\dot{\gamma}$ in the rheological equation of the fluid, i.e.

$$\dot{\gamma} = f(\tau)$$

The interpretation of the data differs for each class of fluid and these will now be treated in turn.

Newtonian fluid

It is also shown in Appendix 2 that for a Newtonian fluid the general functional relation given in Eqn. [2.2.1] simplifies to

$$\mu = (r_2^2 - r_1^2)G/4\pi\, r_1^2\, r_2^2\, \Omega \qquad [2.2.2]$$

This equation implies that a plot of G against Ω is a straight line through the origin, the slope of which gives the viscosity. Hence the viscosity of a Newtonian liquid may be determined in principle by a single experiment. Further experiments at different speeds only serve to improve the accuracy of the determination.

Pseudoplastic and dilatant fluids

Eqn. [2.2.2] may also be used to define an apparent viscosity of a non-Newtonian fluid at a given shear rate if the annular gap is sufficiently small for the shear rate to be virtually constant throughout the sample. For instance if the radii of the inner and outer cylinders differ by not more than 5 per cent the

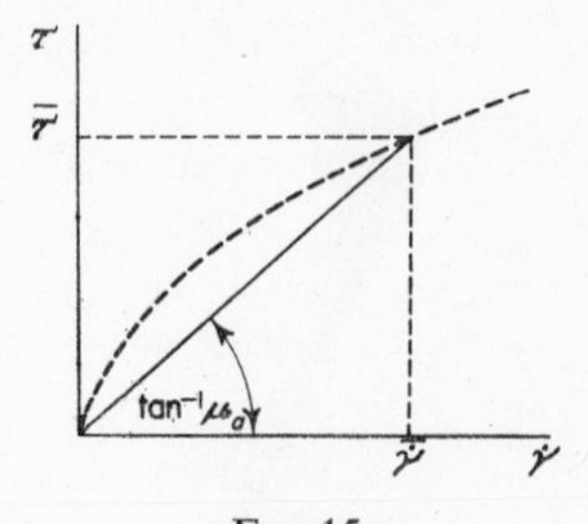

FIG. 15.

variation in both stress and rate of shear will be about 10 per cent (for a Newtonian liquid) and we could say that the viscosity as calculated from Eqn. [2.2.2] will be the apparent viscosity, μ_a, of the non-Newtonian fluid at the shear stress represented by the mean of the stresses at the surfaces of the two cylinders. The flow curve could then be simply constructed as in Fig. 15.

The mean rate of shear and the mean shear stress are conveniently calculated at $r = \sqrt{(r_1 r_2)}$, i.e. at the geometric mean radius. These are given by

$$\bar{\dot{\gamma}} = 2r_1 r_2 \Omega/(r_2^2 - r_1^2) \qquad [2.2.3]$$

and

$$\bar{\tau} = G/2\pi r_1 r_2 \qquad [2.2.4]$$

whence

$$\bar{\tau}/\bar{\dot{\gamma}} = \mu_a = (r_2^2 - r_1^2)G/4\pi\, r_1^2 r_2^2 \Omega \qquad [2.2.5]$$

This method should be used with caution because although it is a simple matter to estimate the variation in rate of shear across the gap for a given ratio of r_2/r_1 for a Newtonian liquid, the corresponding variation for a non-Newtonian fluid in the same apparatus could be much greater. This effect has been pointed out by Wratten.[12]

For a power-law fluid it is shown in Appendix 2 that the apparent viscosity at any radius r is given by

$$\mu_a(r) = \frac{(r_2^2 - r_1^2)G}{4\pi r_1^2 r_2^2 \Omega}\left[n\left\{\frac{1 - (r_1/r_2)^{2/n}}{1 - (r_1/r_2)^2}\right\}\left(\frac{r}{r_1}\right)^{2/n - 2}\right] \qquad [2.2.6]$$

The term inside the square bracket approaches unity as r_1/r_2 approaches unity. This is an accurate equation and should be used if any doubt exists about the previous method.

Anomalous behaviour near the walls can be detected by repeating the experiments with slightly different annular gaps (but keeping these small). If a wall effect is present the curves for different gaps will not coincide.

However, if r_1 and r_2 are not nearly equal the above procedure would be inaccurate and the true flow curve should then be determined by calculating the shear stress and the shear rate at corresponding points in the fluid. The most convenient point for this is at the surface of the inner cylinder. At this point the shear stress is given by

$$\tau_1 = G/2\pi r_1^2 \qquad [2.2.7]$$

Krieger and Maron[13] have shown that the rate of shear at the surface of the inner cylinder is given by an infinite series which converges rapidly when the ratio of r_2/r_1 is less than 1·2. For values of r_2/r_1 less than 1·2 the expression for the shear rate at the inner cylinder reduces to

$$\frac{4\,\pi\Omega a^2}{a^2-1}\left[1+k_1\left(\frac{1}{S}-1\right)+k_2\left(\frac{1}{S}-1\right)^2\right] \qquad [2.2.8]$$

where a, k_1 and k_2 are instrument constants given by

$$a = r_2/r_1$$

$$k_1 = \frac{a^2-1}{2a^2}\left(1+\frac{2}{3}\ln a\right) \qquad [2.2.9]$$

$$k_2 = \frac{a^2-1}{6a^2}\ln a \qquad [2.2.10]$$

and in any experiment S is the slope of the logarithmic plot of torque against rotational speed, Ω, at the particular value of Ω used in the experiment.

Hence the shear stress and the corresponding rate of shear can be found from Eqns. [2.2.7] and [2.2.8] respectively and the flow curve then follows directly by repeating the experiment over a range of speeds.

Bingham plastics

A quantitative picture of the flow pattern when a Bingham plastic is contained in a coaxial cylinder viscometer can be obtained by considering the system as the speed is increased from zero. If the couple per unit height of liquid on the inner cylinder (stationary) is G the shear stresses at the inner and outer walls will be given by τ_1 and τ_2 where

$$\tau_1 = G/2\pi\, r_1^2 \qquad [2.2.11]$$

and

$$\tau_2 = G/2\pi\, r_2^2 \qquad [2.2.12]$$

This means that the shear stress on the inner cylinder, τ_1, is always greater than that on the outer cylinder, τ_2.

Initially if a small couple is applied to the outer cylinder the shear stress, τ_1, will be less than τ_y, the yield stress of the Bingham plastic. Since $\tau_1 > \tau_2$ we have that τ_2 is also less than τ_y and no flow will occur anywhere. Hence the outer cylinder will not move provided that the shear stress on the inner cylinder is less than the yield value, i.e. provided that

$$G/2\pi\, r_1^2 < \tau_y \qquad [2.2.13]$$

If the couple G is now increased so that $\tau_1 > \tau_y > \tau_2$, flow will occur near the inner cylinder only. The yield stress, τ_y, will now be attained at a radius r_y where $r_1 < r_y < r_2$ given by

$$G/2\pi\, r_y^2 = \tau_y \qquad [2.2.14]$$

Now the material contained between r_1 and r_y will flow while the material between r_y and r_2 (where the stress in the material is still less than the yield stress) will remain solid and will move as a rigid body attached to the outer cylinder.

If the couple is increased still further so that $\tau_2 > \tau_y$, i.e.

$$G/2\pi\, r_2^2 > \tau_y \qquad [2.2.15]$$

the speed will increase further and flow will occur throughout the material. In this condition it is shown in Appendix 2 that the relation between the couple per unit height, G, and the speed, Ω, is given by

$$\Omega = \frac{G}{4\pi\,\mu_p}\left(\frac{1}{r_1^2} - \frac{1}{r_2^2}\right) - \frac{\tau_y}{\mu_p}\ln\frac{r_2}{r_1} \qquad [2.2.16]$$

Hence if Ω is plotted against G the curve will become a straight line of slope $(1/r_1^2 - 1/r_2^2)/4\pi\,\mu_p$ when G exceeds the value $2\pi\,\tau_y\, r_2^2$. This is shown in Fig. 16.

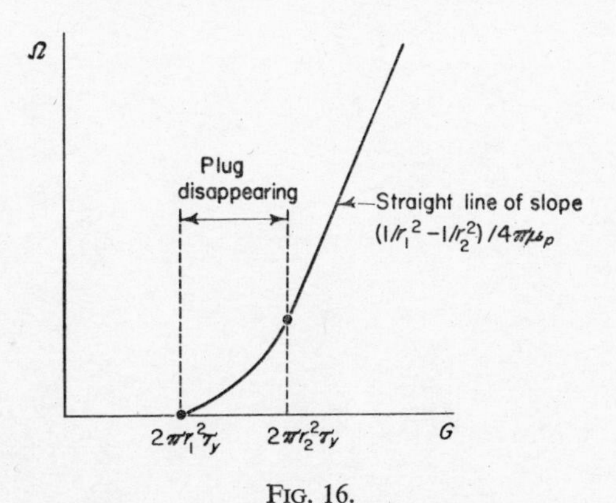

FIG. 16.

An alternative plot is shown in Fig. 17.

These plots show how μ_p and τ_y, the two characteristic quantities of a Bingham plastic, may be evaluated from measurements of speed and applied torque in a coaxial cylinder viscometer.

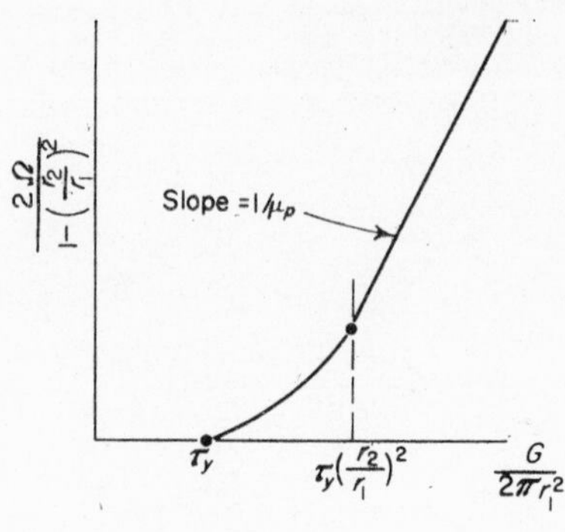

FIG. 17.

(b) Rotating bob in an infinite fluid

This instrument is an extension of the coaxial type for which $r_2 = \infty$, and is shown diagrammatically in Fig. 18.

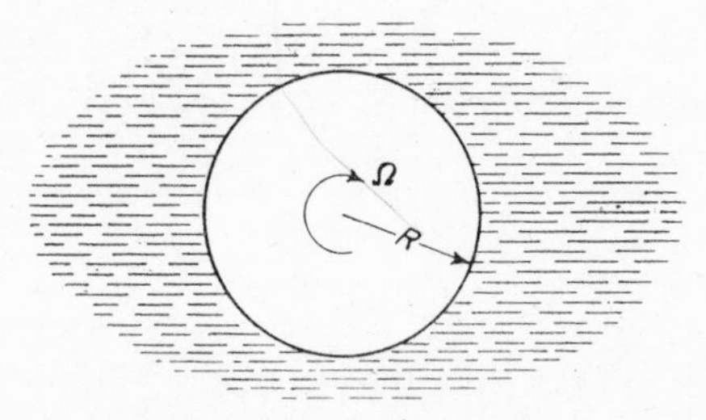

FIG. 18.

The shear stress at the surface of the bob is $G/2\pi R^2$ and the rate of shear at the surface can be shown to be given by [13]

$$4\pi\, \Omega/S \qquad [2.2.17]$$

where S is the slope of the logarithmic plot of torque versus speed.

The shear stress and rate of shear being thus determined at the same point in the fluid, the construction of the flow curve follows simply by measuring the applied torque for different rotational speeds of the bob.

(c) The cone and plate viscometer

This type of instrument, which is shown diagrammatically in Fig. 19, consists of a flat plate and a rotating cone with a very obtuse angle. The apex of the cone just touches the plate surface and the fluid fills the narrow gap formed by the cone and plate. Temperature control is usually applied to the lower plate.

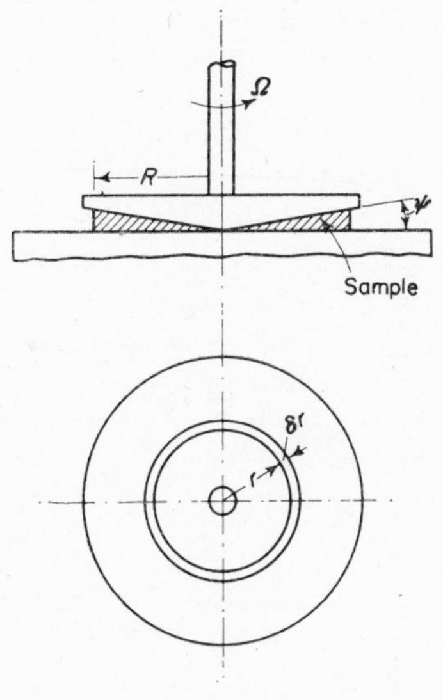

FIG. 19.

If the angle ψ is very small, say less than 0·5°, and the average gap width is correspondingly small, say less than 0·5 mm, the whole sample will be subjected to a constant rate of shear and the end effects will be negligible. This simplifies the analysis of the data for non-Newtonian fluids because it gives the apparent viscosity as a function of shear rate directly as follows.

The linear velocity at radius r is Ωr and the gap width at radius r is $r \tan \psi$. Hence the rate of shear at radius r is given by

$$\dot{\gamma}(r) = \Omega r / r \tan \psi = \Omega \tan \psi$$

This means that the rate of shear is constant throughout the sample and independent of r.

Further

$$\dot{\gamma}(r) = \Omega / \psi \text{ if } \psi \text{ is small.} \qquad [2.2.18]$$

Also since $\dot{\gamma} = f(\tau)$ and $\dot{\gamma}$ is constant we have that τ is constant.

Therefore

$$2\pi \, \tau \int_0^R r^2 \mathrm{d}r = G$$

hence

$$\tau = \frac{3G}{2\pi R^3} \quad [2.2.19]$$

and

$$\mu_a = \frac{3G}{2\pi R^3} \Big/ \frac{\Omega}{\psi} \quad [2.2.10]$$

Eqn. [2.2.10] gives the apparent viscosity at a rate of shear given by Ω/ψ. Alternatively the flow curve could be constructed by plotting the rate of shear (Ω/ψ) against the corresponding shear stress $(3G/2\pi R^3)$ directly.

An instrument of this type has been designed by Piper and Scott[14] and a commercial instrument based on this principle is described in Chapter 6. Mooney and Ewart[15] have used the same principle in the design of the bottom of a coaxial cylinder viscometer, so that the end effect could be calculated directly. Weltmann[16] has also described such an instrument.

2.3 TIME-INDEPENDENT FLUIDS IN CAPILLARY TUBE VISCOMETERS

The characterization of non-Newtonian fluids in capillary tube viscometers entails simultaneous observations of the volume rate of flow, Q, through the tube and the corresponding pressure drop, ΔP, over a given length. This procedure is then repeated to cover a range of Q and ΔP. The apparatus consists essentially of a means of applying a known pressure difference to the tube, which may be approximately 0·05 in. in diameter and 18 in. long, with suitable equipment for the accurate measurement of the rate of flow. Details of suitable instruments and the experimental techniques will be given in Chapter 6.

It has been pointed out by several authors that on the assumptions that

(1) flow is laminar so that each particle moves in a straight line, at constant velocity, parallel to the axis of the tube,
(2) there is no slip at the wall,
(3) the rate of shear at a point depends only on the shearing stress at that point, i.e. $\dot{\gamma} = f(\tau)$ only,

it may be shown that the functional relation connecting $Q/\pi a^3$ and τ_w (the shear stress at the tube wall) will be independent of the tube diameter no matter how complex the connexion between shear stress and rate of shear. This relation, which is derived in Section 3.1. may be written in the form

$$\frac{Q}{\pi a^3} = \frac{1}{\tau_w^3} \int_0^{\tau_w} \tau^2 f(\tau) \, \mathrm{d}\tau \quad [2.3.1]$$

where the function f is the function connecting shear stress and rate of shear.

Since the integral in Eqn. [2.3.1] is a function of its limits only, it is seen that $Q/\pi a^3$ depends only on τ_w no matter how complex $f(\tau)$ may be. The stress at the pipe wall, τ_w, is given by

$$2\pi a L\tau_w = \pi a^2 \Delta P$$

where ΔP is the pressure drop over length L.

Therefore

$$\tau_w = a\Delta P/2L \qquad [2.3.2]$$

If from the observed pressure drop in the tube and the flow rate, $Q/\pi a^3$ is plotted against $a\Delta P/2L$ for different tube diameters on the same diagram, all the data should lie on a single curve if the three assumptions listed above are valid. If the possibility of non-laminar flow can be ruled out and the fluid is known to be time-independent then a separation of the curves for different tube diameters can be interpreted as evidence of anomalous flow behaviour or effective slip near the tube wall and this situation can be analysed on the basis of a theoretical relation which has been given by Oldroyd [17], discussed in Appendix 1.

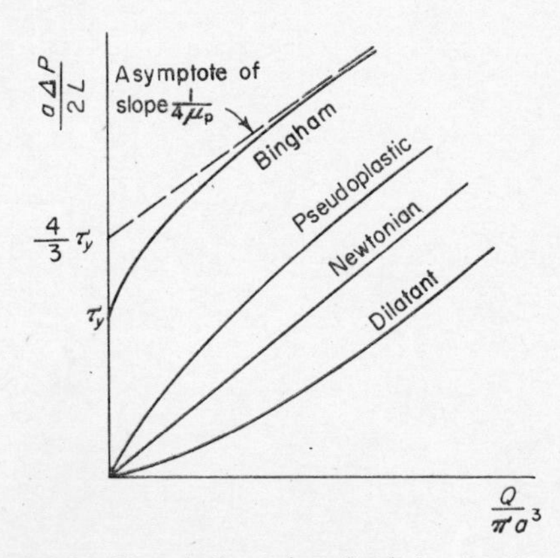

FIG. 20. Time-independent fluids in pipe flow.

The graph of $Q/\pi a^3$ versus $a\Delta P/2L$ may be used to derive the relation between shear stress and rate of shear as follows.

We have that

$$\frac{Q}{\pi a^3} = F(\tau_w) = \frac{1}{\tau_w^3}\int_0^{\tau_w} \tau^2 f(\tau)\, d\tau \qquad [2.3.3]$$

Hence it follows that

$$f(\tau_w) = \frac{1}{\tau_w^2} \frac{\mathrm{d}\,[\tau_w^3 F(\tau_w)]}{\mathrm{d}\,\tau_w} \qquad [2.3.4]$$

The function F is known graphically, enabling $f(\tau_w)$, the rate of shear at the tube wall, to be calculated for any value of τ_w. The graph of τ_w against $f(\tau_w)$ is, of course, the flow curve. This procedure is sound in principle but in practice the curve for $F(\tau_w)$ is often not sufficiently sharply defined to allow the differentiation involved in Eqn. [2.3.4] to be carried out accurately.

2.4 GENERALIZED CLASSIFICATION FROM TUBE VISCOMETER DATA

Rabinowitsch[53] and Mooney[18] have developed an expression for the rate of shear at the wall of a tube which is independent of fluid properties provided that the fluid is not time-dependent. The expression takes the form

$$-\left(\frac{\mathrm{d}u}{\mathrm{d}r}\right)_w = 3\left(\frac{Q}{\pi a^3}\right) + \frac{a\Delta P}{2L} \frac{\mathrm{d}\,(Q/\pi a^3)}{\mathrm{d}\,(a\Delta P/2L)} \qquad [2.4.1]$$

Introducing the mean velocity, u_m, which is equal to $Q/\pi a^2$ and putting $D = 2a$, Eqn. [2.4.1] can be rearranged to give

$$-\left(\frac{\mathrm{d}u}{\mathrm{d}r}\right)_w = \frac{3}{4}\left(\frac{8u_m}{D}\right) + \frac{1}{4}\left(\frac{8u_m}{D}\right) \frac{\mathrm{d}\ln\,(8u_m/D)}{\mathrm{d}\ln\,(D\Delta P/4L)} \qquad [2.4.2]$$

If we denote the derivative in Eqn. [2.4.2] by $1/n'$ we can write this equation as

$$-\left(\frac{\mathrm{d}u}{\mathrm{d}r}\right)_w = \frac{3n'+1}{4n'} \frac{8u_m}{D} \qquad [2.4.3]$$

This is the most convenient form of Eqn. [2.4.1] and it was first proposed by Metzner and Reed.[19]

From the definition of n', namely,

$$n' = \frac{\mathrm{d}\ln\,(D\Delta P/4L)}{\mathrm{d}\ln\,(8\,u_m/D)} \qquad [2.4.4]$$

it is easily seen that we can write

$$\tau_w = D\Delta P/4L = k'(8u_m/D)^{n'} \qquad [2.4.5]$$

This is the equation of the tangent to the logarithmic plot of τ_w and $8u_m/D$ at the chosen value of τ_w, and k' corresponds to a particular value of τ_w. From the logarithmic plot of $D\Delta P/4L$ against $8u_m/D$ we can find n' and $8u_m/D$ at a selected value of the shear stress at the wall $D\Delta P/4L$. The corresponding rate of shear at the wall then follows from Eqn. [2.4.3]. In this way we can determine the shear stress and rate of shear at corresponding points, and the flow curve can be drawn by repeating this procedure at different values of the wall shear stress. If the slope of the logarithmic plot is constant over a range of shear stresses, Eqn. [2.4.5] represents the equation of this straight line. If the plot is not a straight line, n' and k' must be evaluated at the appropriate shear stress.

This approach is superficially similar to the power law which for pipe flow may be written

$$\tau = k\,(\mathrm{d}u/\mathrm{d}r)^n \tag{2.4.6}$$

and k' and n' are closely related to k and n. k' is a measure of the consistency of the material and is known as the '*consistency index*'. The larger k' is the more viscous the fluid. n' is that physical property of a fluid which characterizes its degree of non-Newtonian behaviour, and the greater the divergence of n' from unity (greater than unity for dilatant fluids and less for pseudoplastics) the more non-Newtonian is the fluid; n' is known as the '*flow-behaviour index*.' However, although Eqns. [2.4.5] and [2.4.6] are similar, the former has some important advantages for engineering applications. It is, in fact, a direct relation between the pressure drop, ΔP, and the flow rate (or average velocity of the fluid, u_m) in terms of the tube dimensions and the characteristic parameters of the fluid k' and n'. Hence it may be used for a rigorous pipeline design provided k' and n' are known at the appropriate values of $8u_m/D$ under consideration. However if the power law (Eqn. [2.4.6] is used for pipeline design it must be integrated, as will be shown in the next chapter, and this implies that the index n is constant over the whole range of shear stresses encountered in the pipe, i.e. from $D\Delta P/4L$ at the wall to zero at the centre. This is frequently not the case.

Another important point to note is that this approach is applicable to all fluids (including Bingham plastics) and not merely to fluids of the power law type.

The relation between n and n' at any particular shear stress may be derived by taking logarithms of Eqn. [2.4.3], differentiating and dividing by $\mathrm{d}(\ln \tau_w)$.

This gives

$$\frac{\mathrm{d}\,[\ln(-\mathrm{d}u/\mathrm{d}r)_w]}{\mathrm{d}(\ln \tau_w)} = \frac{\mathrm{d}(\ln 8u_m/D)}{\mathrm{d}(\ln \tau_w)} + \frac{\mathrm{d}\left(\ln \dfrac{3n'+1}{4n'}\right)}{\mathrm{d}\ln \tau_w} \tag{2.4.7}$$

Also from Eqn. [2.4.6] we have that

$$\ln \tau = \ln k + n \ln (-\mathrm{d}u/\mathrm{d}r) \qquad [2.4.8]$$

Hence by differentiation, noting that k is constant, we have

$$\frac{\mathrm{d}(\ln \tau)}{\mathrm{d}[\ln (-\mathrm{d}u/\mathrm{d}r)]} = n \qquad [2.4.9]$$

(It should be noted here that a comparison of Eqns. [2.4.9] and [2.4.4] brings out the distinction between n and n' : n is the slope of the logarithmic plot of shear stress versus the rate of shear, whereas n' is the slope of the logarithmic plot of shear stress at the pipe wall versus the quantity $8u_m/D$ and the latter is not equal to the rate of shear at the wall except for the special case of a Newtonian fluid.)

Substituting from Eqns. [2.4.4] and [2.4.9] into [2.4.7] we obtain

$$\frac{1}{n} = \frac{1}{n'} + \frac{\mathrm{d} \ln \left(\dfrac{3n'+1}{4n'}\right)}{\mathrm{d} \ln \tau_w}.$$

Rearranging gives

$$n = \frac{n'}{1 - \dfrac{1}{3n'+1}\left(\dfrac{\mathrm{d}\, n'}{\mathrm{d} \ln \tau}\right)} \qquad [2.4.10]$$

where n' now refers to the shear stress τ.

It is seen from Eqn. [2.4.10] that if n' does not vary with shear stress, i.e. if the logarithmic plot of $D\Delta P/4L$ and $8u_m/D$ is a straight line, the derivative in the denominator is zero and n becomes equal to n'. In general, however, this is not so. n can always be determined from n' but the converse is not true.

A relation may also be derived between k and k'. By integration of Eqn. [2.4.6] it can be shown (see Section 3.1) that

$$\frac{D\Delta P}{4L} = k\left(\frac{3n+1}{4n}\right)^n \left(\frac{8u_m}{D}\right)^n. \qquad [2.4.11]$$

Therefore by comparing Eqns. [2.4.11] and [2.4.5] we see that

$$k' = k\left(\frac{3n+1}{4n}\right)^n \qquad [2.4.12]$$

and, further, $k' = k$ when $n = 1$, i.e. when the fluid is Newtonian. In general k' and k are different.

2.5 METHODS AVAILABLE FOR THE CHARACTERIZATION OF TIME-DEPENDENT FLUIDS

(a) Capillary-tube viscometers

It has been stated previously that the curve of $Q/\pi a^3$ (or $8u_m/D$) against $D\Delta P/4L$ is unique and independent of tube dimensions only if the fluid properties are independent of time. Thixotropic fluids show decreases in consistency with increasing times of shear at a given shearing stress, and if capillary-tube data for these fluids were plotted in this way the result would be as in Fig. 21.

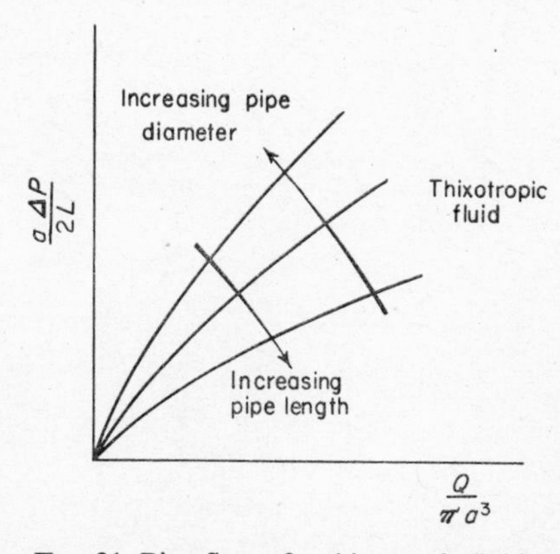

FIG. 21. Pipe flow of a thixotropic fluid.

More thixotropic breakdown would be indicated with longer pipes and smaller diameters, i.e. longer times of shear and higher rates of shear, respectively. This effect has been demonstrated experimentally by Ambrose and Loomis.[20] The reverse would be true of materials which show a structure build-up on shearing (the so-called rheopectic fluids) and at a given value of $D\Delta P/4L$ the value of $8u_m/D$ would increase with increasing tube diameter and decrease with increasing tube length. For the purposes of characterizing time-dependent fluids quantitatively this method is unsatisfactory and it does appear that rotational instruments, which enable the measurements to be made under precise and uniform rates of shear, offer definite advantages.

(b) Rotational viscometers

In a rotational viscometer, such as the Couette coaxial type, thixotropic breakdown would be indicated by a progressive decrease in torque at a constant rotational speed until the position of equilibrium breakdown (dependent on the speed) was reached.

In order to determine the extent and nature of thixotropic breakdown it is necessary to plot at least two curves, the 'up' and 'down' curves, obtained by gradually increasing the rate of shear to any desired maximum value and then decreasing it again to zero. Fig. 22 shows the possible results of such an

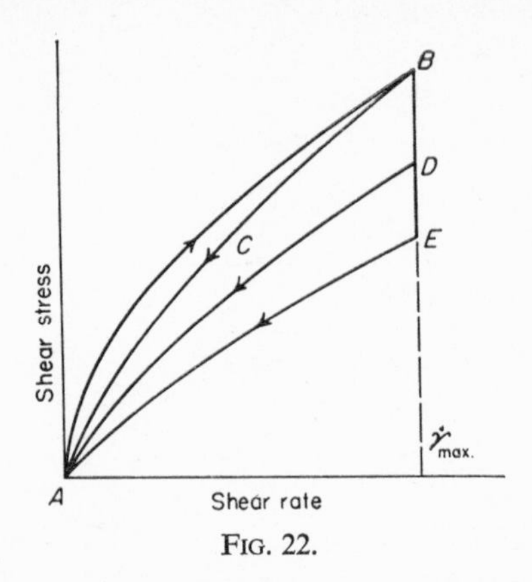

FIG. 22.

experiment. The 'up curve' AB is determined by increasing the rate of shear at a constant rate to $\dot{\gamma}_{max}$. If the 'down curve' is started immediately it would follow a path such as BCA in general. If the rate of shear is held at $\dot{\gamma}_{max}$ for a time the shear stress would fall off due to thixotropic breakdown to some value represented by, say, D. If the 'down curve' were then started it would follow path DA. If the shear rate were held at $\dot{\gamma}_{max}$ for an infinite time the stress would drop to some point E and the 'equilibrium down curve', EA, would result corresponding to maximum thixotropic breakdown.

Green [21] and Green and Weltmann [22] have suggested methods for the quantitative characterization of thixotropic materials on the basis of the shape and area of hysteresis loops such as ABD. Similar considerations would presumably apply to rheopectic materials but, so far as the author is aware, no experimental measurements similar to those described above for thixotropic materials have been reported.

Green and Weltmann characterized thixotropic materials of the Bingham plastic type in terms of two coefficients, B and M. The *coefficient of thixotropic breakdown with time*, B, indicates the rate of breakdown with time at constant shear rate, and M is the *coefficient of thixotropic breakdown due to increasing shear rate*. These coefficients can be obtained from experiments in a rotational viscometer as follows. To determine B the viscometer is set to give a predetermined maximum speed and when this is reached, say point A in Fig. 22(a) the material is sheared for a time t_1 when the torque will fall to some value such as B_1. The down curve is then begun. The material is rested and the experiment repeated but now the material is sheared for a longer time at the maximum speed, say t_2, so that the torque drops to B_2 before the down curve is begun. Let μ_{p1} and u_{p2} be the two plastic viscosities corresponding to the two down curves, i.e. the slopes of $B_1 0$ and $B_2 0$ respectively. Weltmann

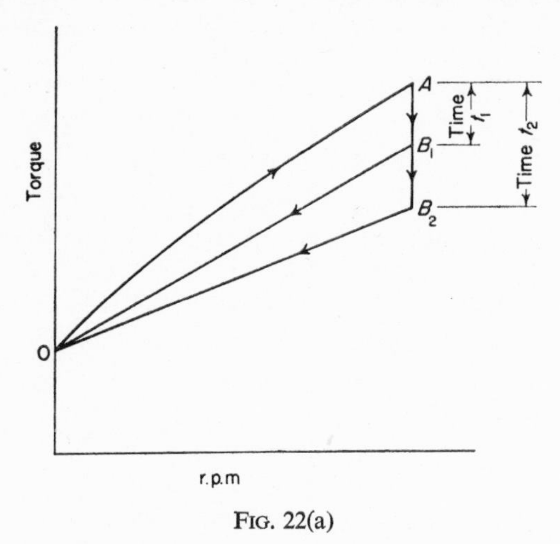

FIG. 22(a)

has shown that the plastic viscosity decreases with the logarithm of the time and the coefficient B is defined by

$$B = -t \, d\mu_p/dt = \text{constant}$$

whence from the two experiments we obtain

$$B = (\mu_{p1} - \mu_{p2})/ \log (t_2/t_1)$$

In order to determine M, two consecutive hysteresis loops are plotted with different maximum speeds, say N_1 and N_2 as in Fig. 22(*b*).

Weltman has shown that breakdown is proportional to the top rate of shear and that

$$\exp (-2\mu_p/M) \propto N^2$$

Therefore if the plastic viscosities corresponding to the two down curves $A_1 0$ and $A_2 0$ are now μ_{p1} and μ_{p2} we have that

$$M = (\mu_{p1} - \mu_{p2})/ \log (N_2/N_1)$$

The physical significance of M is that it is the loss in shear stress per unit increase in rate of shear.

This approach has been used to characterize the thixotropic behaviour of pigment-vehicle suspensions. There is, however, controversy as to whether such methods have any real theoretical significance.

In summary, the quantitative characterization of time-dependent fluids is difficult and, as yet, reliable experimental methods have not been developed.

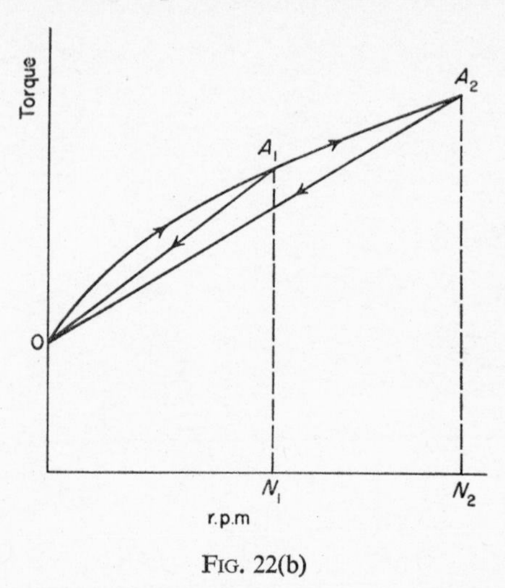

FIG. 22(b)

For industrial applications this is not a serious drawback at the present time because these time-dependent fluids are encountered far less frequently than time-independent fluids, particularly those of the pseudoplastic type.

2.6 EXPERIMENTAL CHARACTERIZATION OF VISCOELASTIC MATERIALS

Methods available

In principle there are two main methods of specifying viscoelastic materials. These are:

(1) by specifying the constants occurring in the differential equation relating stress and strain, i.e. the constants $\alpha_0 \ldots \alpha_n$ and $\beta_0 \ldots \beta_n$ in the equation of state

$$\sum_{n=0}^{N} \alpha_n \mathrm{D}^n \tau = \sum_{m=0}^{M} \alpha_m D^m \gamma \qquad [2.6.1]$$

(2) by stating the response curve of the material to some specified stress or strain. Any time function of stress or strain would suffice in principle but, as in the case of automatic control theory, step and sinusoidal inputs are the easiest to manage mathematically.

These two methods are of course equivalent and in the case of linear viscoelastic materials it is possible to convert one into the other mathematically. A linear material is one which obeys the principle of superposition. Consider for instance a creep experiment. If $\gamma_1(t)$ is the strain at time t when the material is subjected to a stress τ_1, and $\gamma_2(t)$ is the strain at the same time if the material is subjected to a stress τ_2, then if the material is linear, the strain at time t when subjected to stress $\tau_1 + \tau_2$ will be $\gamma_1(t) + \gamma_2(t)$. Linear materials only will be considered in the following analysis.

Experimentally the rheological properties are characterized by method (2), and we can divide this method into two sections according to the type of experiment adopted.

2.7 TRANSIENT EXPERIMENTS—STEP RESPONSE

(a) Strain retardation or creep experiments

The creep function, Eqn. [1.5.10] is perhaps the most convenient method of specifying the properties of a viscoelastic material which has predominantly 'solid' properties, i.e. a material which is capable of supporting its own weight without appreciable distortion as compared with a 'liquid' which flows and must be confined.

The creep function has been defined as $\phi(t)$ where

$$\phi(t) = \gamma(t)/\tau_0 \qquad [2.7.1]$$

i.e. $\gamma(t)$ is observed experimentally following imposition of stress τ_0 at $t = 0$.

A typical creep diagram is shown in Fig. 23.

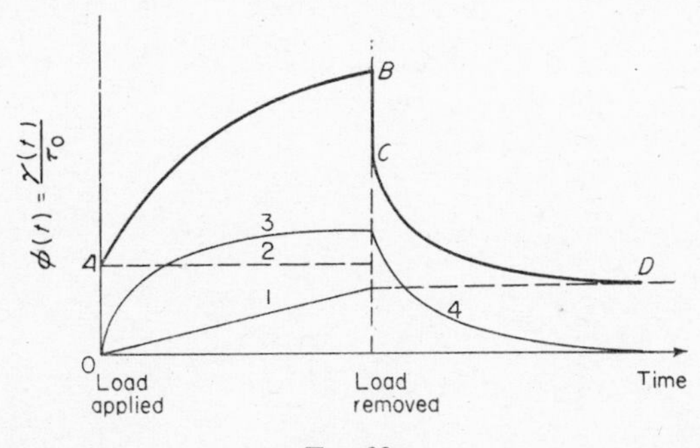

FIG. 23.

In general the curve consists of three parts:

(1) viscous flow

(2) instantaneous elastic strain

(3) delayed elastic strain.

These are the curves 1, 2, 3 in Fig. 23. The creep curve (curve AB) is the sum of 1, 2 and 3. This treatment is justified for a linear material which obeys the principle of superposition. If the experiment is continued for a sufficiently long time the retarded elastic strain curve will fall off to a zero slope and the creep curve will then have the same slope as the viscous flow component.

Suppose at this point that the load is removed. The elastic strain in the material will recover immediately, accounting for the vertical line BC which will be equal to 0A. The delayed elastic strain will also completely recover in time but the viscous deformation will remain. The recovery curve CD will therefore be a mirror image of curves 2 plus 3. The delayed elastic component is shown separately as curve 4. This is a mirror image of curve 3.

The equivalent model which would give a creep curve such as this could be a generalized Voigt body with one element having a dash-pot of zero viscosity and another having a spring of zero stiffness. This is represented diagrammatically in Fig. 24 and the rheological equation would be of the form

$$\phi(t) = \frac{\gamma(t)}{\tau_0} = J + \frac{t}{\mu} + \sum_{n=0}^{N} J_n \left[1 - \exp\left(-t/\lambda_n\right)\right]$$

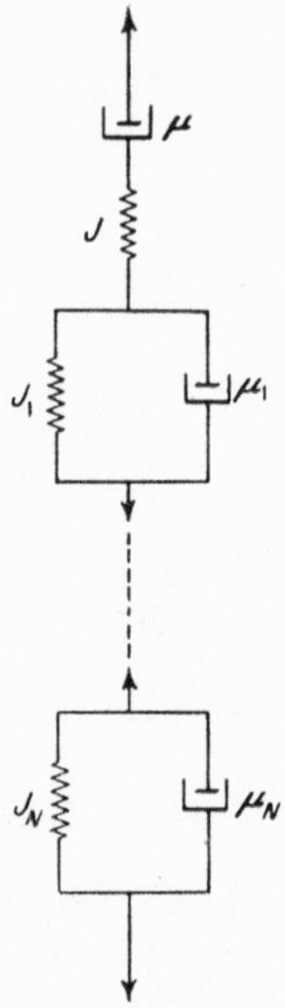

FIG. 24.

This equation would also be obeyed by a Voigt body in series with a simple Maxwell element.

The details of the experimental techniques used for the determination of creep curves have been fully discussed by Leaderman.[23] 'Solid' samples are usually studied in extension. 'Liquid' samples are studied in shear between rotating coaxial cylinders or between a rotating cone and plate.

Ideally the creep curve OAB should be determined over the complete time range 0 to ∞. Experimentally it is very difficult to determine the short-time response accurately. This means that the short-time behaviour of the material will not show up. In terms of models we could say that a Voigt element with a retardation time less than the smallest time observable experimentally (say less than one second) would behave as a pure spring. In general the delayed elastic character of an element can only be indicated by experiments on a time scale approximately equal to the retardation time.

Consequently the transient creep function is only useful at the long end of the time scale.

(b) Stress relaxation function

Experimentally the stress relaxation curve is determined by suddenly imposing a strain on the material at time zero, and then measuring the stress to maintain this strain as a function of time. The relaxation function is then given by

$$\psi(t) = \tau(t)/\gamma_0 \qquad [2.7.2]$$

This technique is most useful for viscoelastic 'liquids' which flow and must be confined.

In general the relaxation curve would be as indicated in Fig. 25.

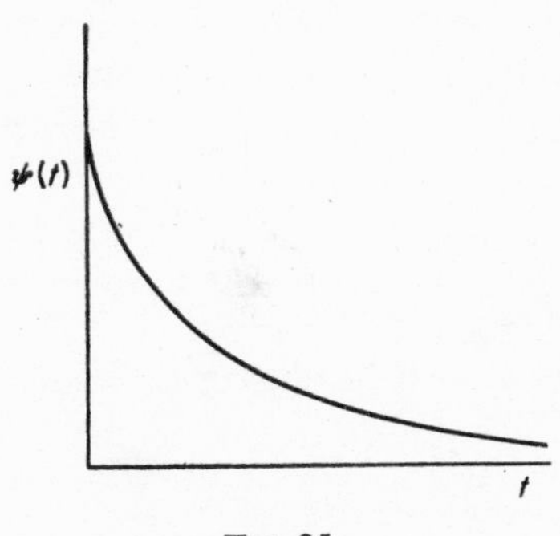

FIG. 25.

Although this method is most used for 'liquids' it has been applied to 'solids' in extension. 'Liquids' are usually studied in shear between coaxial cylinders or a cone and plate.

As in the case of creep experiments relaxation methods are best suited to the long end of the time scale. There is another experimental difficulty here, however, since ideally the strain should be imposed in zero time. The effect of a finite rate of application of strain is shown in Fig. 26, and this effect can often lead to spurious conclusions.

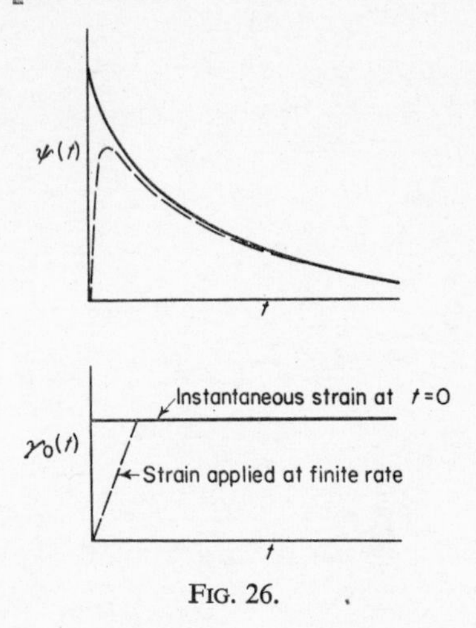

FIG. 26.

2.8 DYNAMIC EXPERIMENTS—FREQUENCY RESPONSE

In the previous section step-function experiments were considered and it was seen that their main limitation is that the short-time behaviour of the system does not show up clearly. This same limitation is encountered when step functions are used to study the dynamic characteristics of process plants in automatic control problems. In this case the dynamic characteristics of the plant can be more accurately determined by means of a sinusoidal forcing signal. This technique, the so-called frequency response method, is also used to study the short-time behaviour of viscoelastic materials.

If a sinusoidal stress is applied to a linear viscoelastic material the strain will also vary sinusoidally with time but it will be out of phase with the applied stress. This is illustrated in Fig. 27.

At any given frequency the behaviour of the material must be specified by two experimentally determined quantities. These may be chosen in various ways. For instance we could specify the ratio of the amplitudes of stress and the resulting strain, τ_0/γ_0, and the phase difference between them, ϕ. These terms would then be analogous to the quantities attenuation and phase lag used in the frequency response of process plants. These quantities would of course depend on the frequency.

However, in the dynamic analysis of viscoelastic materials it has become customary to separate the stress (which is not in phase with the strain) into two components, one which is in phase with the strain and one which is 90° out of phase. Alternatively the strain is separated into two components, one in phase and the other 90° out of phase with the stress. These schemes are illustrated in Fig. 28.

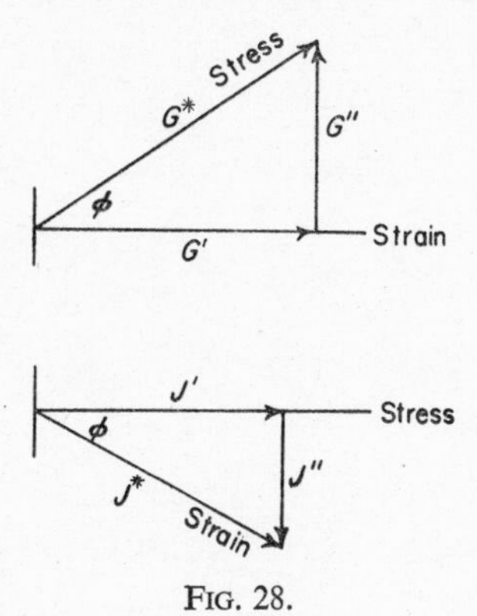

FIG. 28.

The component in phase with the stress, G', is defined as

$$G'(\omega) = \frac{\text{component of stress in phase with strain}}{\text{strain}}$$

and similarly

$$G''(\omega) = \frac{\text{component of stress } 90^\circ \text{ out of phase with strain}}{\text{strain}}$$

Likewise if we adopt the second scheme we define

$$J'(\omega) = \frac{\text{component of strain in phase with stress}}{\text{stress}}$$

and

$$J''(\omega) = \frac{\text{component of strain } 90^\circ \text{ out of phase with stress}}{\text{stress}}$$

The resultants in each case, $G^*(\omega)$ and $J^*(\omega)$, are given by

$$G^*(\omega) = G'(\omega) + i\,G''(\omega)$$

and [2.8.1]

$$J^*(\omega) = J'(\omega) + iJ''(\omega)$$

also

$$\tan \phi = G''/G' = J''/J'$$

The two schemes are of course equivalent, and either $J^*(\omega)$ or $G^*(\omega)$ define the rheological properties of the material and $J^*(\omega)$ is simply the reciprocal of $G^*(\omega)$, i.e. $|J^*| = 1/|G^*|$. However $J^*(\omega)$ is more convenient in describing viscoelastic 'solids' and $G^*(\omega)$ is preferred for 'liquids'.

2.9 DYNAMIC RESPONSE OF THE VOIGT AND MAXWELL BODIES

Consider the equation of a Maxwell body

$$\dot{\gamma}(t) = \tau(t)/\mu + \dot{\tau}(t)/G$$

We can write this

$$(1 + \lambda\,\mathrm{D})\,\tau(t) = \mu\,\mathrm{D}\,\gamma(t)$$

where

$$\mathrm{D} = \frac{\mathrm{d}}{\mathrm{d}t} \text{ and } \lambda = \mu/G$$

Suppose that the strain is varied sinusoidally, i.e.

$$\gamma(t) = \gamma_0 \sin \omega t$$

Then

$$\mathrm{D}\gamma(t) = \gamma_0\omega \cos \omega t$$

and we have

$$\tau(t) = \frac{\mu\gamma_0\omega \cos \omega t}{(1 + \lambda\mathrm{D})}$$

The solution of this equation may be written

$$\tau(t) = \frac{\mu\gamma_0\omega}{(1 + \omega^2\lambda^2)^{\frac{1}{2}}} \sin\left(\omega t + \tan^{-1}\frac{1}{\omega\lambda}\right) \qquad [2.9.1]$$

from which it is seen that the resulting stress is also sinusoidal, of the same frequency as the strain but out of phase with it.

Further we have that

$$G^*(\omega) = G'(\omega) + iG''(\omega) = \tau/\gamma$$

i.e.

$$G'(\omega) + iG''(\omega) = \frac{\mu\omega}{(1 + \omega^2\lambda^2)^{\frac{1}{2}}} \frac{\sin\left(\omega t + \tan^{-1}\dfrac{1}{\omega\lambda}\right)}{\sin \omega t} \quad [2.9.2]$$

The right-hand side of Eqn. [2.9.2] may be separated into real and imaginary parts to give

$$G'(\omega) = G\omega^2\lambda^2/(1 + \omega^2\lambda^2) \quad [2.9.3]$$

and

$$G''(\omega) = G\omega\lambda/(1 + \omega^2\lambda^2) \quad [2.9.4]$$

These give the components of G^* for a single Maxwell element. It is easily seen from this that for a generalized Maxwell body described by Eqn. [1.5.15] we would have similarly

$$G'(\omega) = \int_0^\infty \frac{\omega^2\lambda^2 G(\lambda)}{(1 + \omega^2\lambda^2)} \, d\lambda \quad [2.9.5]$$

and

$$G''(\omega) = \int_0^\infty \frac{\omega\lambda G(\lambda)}{(1 + \omega^2\lambda^2)} \, d\lambda \quad [2.9.6]$$

It is easily shown by similar reasoning for a Voigt element whose equation is

$$\tau/G = \gamma + \lambda\dot{\gamma} = J\tau$$

that

$$J'(\omega) = J/(1 + \omega^2\lambda^2) \quad [2.9.7]$$

and

$$J''(\omega) = \omega\lambda J/(1 + \omega^2\lambda^2) \quad [2.9.8]$$

where

$$J^* = \gamma/\tau = J' + i J''$$

Similarly for the generalized Voigt body we have

$$J'(\omega) = \int_0^\infty \frac{J(\lambda)}{(1 + \omega^2\lambda^2)} \, d\lambda \quad [2.9.9]$$

and

$$J''(\omega) = \int_0^\infty \frac{\omega\lambda J(\lambda)}{(1 + \omega^2\lambda^2)} \, d\lambda \quad [2.9.10]$$

It should be noted here that $J(\lambda)$ may be determined from the experimentally measured $J'(\omega)$ or alternatively from $J''(\omega)$ from Eqns. [2.9.9] and [2.9.10]. Similar $G(\lambda)$ can be found from $G'(\omega)$ or $G''(\omega)$. These transformations are discussed later in Section 2.11.

2.10 EXPERIMENTAL TECHNIQUES OF FREQUENCY RESPONSE ANALYSIS

The apparatus and experimental techniques used to measure the properties of viscoelastic materials, i.e. the determination of $G^*(\omega)$ for viscoelastic liquids and $J^*(\omega)$ for viscoelastic solids, are varied and rather complex. A full description of these methods is outside the scope of this book and the reader is referred to the excellent review by Ferry[(11)] for the details of the various techniques and a critical comparison of them. All that will be given here is an outline of the principles involved in some of the more important methods.

The methods can be divided into two groups depending on whether the characteristic dimension of the sample (for example the annular gap in a coaxial viscometer) is small or large compared with the wavelength of elastic waves propagated at the frequency of measurement. If this dimension is small compared with the wavelength of elastic waves the inertia of the sample can be neglected, but if it is large viscoelastic waves will be propagated and the inertia of the sample will be significant.

Methods designed primarily for measurements on 'liquids' rather than 'solids' will be described since it is felt that the former are more important in process engineering applications.

(a) Inertia of sample insignificant

Two methods fall readily into this group.

(1) *Direct measurements of stress and strain*

In this method the sample is subjected to sinusoidal deformation and both the stress and strain are recorded directly as functions of time by suitable means. The in-phase and out-of-phase components are then easily determined.

Markovitz *et al.*[(25)] have described a small coaxial cylinder type of instrument based on this principle which is designed to study viscoelastic liquids. One of the cylinders is oscillated sinusoidally and the stress and strain are measured electrically.

The frequency range covered by this method is from about 10^{-6} to 10^{2} c/s and it is continuously variable.

(2) *Resonance devices with added inertia*

If the applied sinusoidal force operates through a mass whose inertia is large compared with the inertia of the sample (for instance a large coaxlia cylinder instrument) the dynamic properties of the sample can only be determined at the resonant frequency of the apparatus. This type of apparatus is usually simpler than those designed to measure stress and strain directly but the frequency of measurement is restricted and can only be varied by changing the inertia, and hence the resonant frequency, of the apparatus. The range of frequency for this type is about 10^{-2} to 10^5 c/s.

A coaxial instrument of this type has been described by Van Wazer and Goldberg.[26]

(b) Inertia of the sample significant

Two methods in this group are particularly convenient.

(1) *Wave propagation*

When the size of the sample is large compared with the wavelength of viscoelastic waves these waves will be propagated and the measurement of their wavelength and attenuation provides the basis of this method of measuring the viscoelastic properties of the sample. This type of instrument has been described by Ferry.[24]

Consider for example a shear wave generated in a large sample by oscillating a thin plate in its own plane as in Fig. 29.

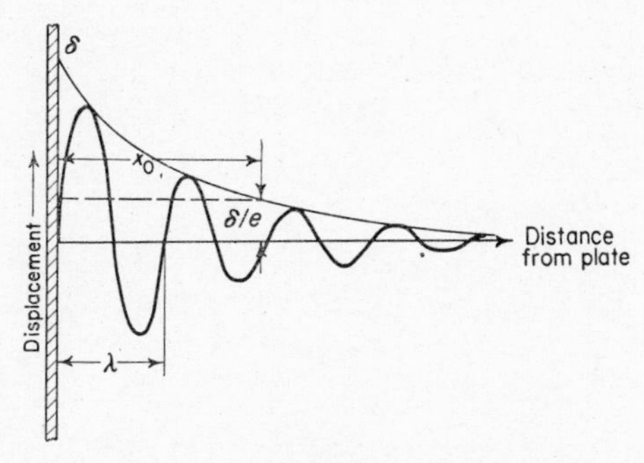

FIG. 29.

The components of the complex modulus G^* may be found from[24]

$$G'(\omega) = \frac{\omega^2\lambda^2\,\rho\,[4\pi^2 - (\lambda/x_0)^2]}{[4\pi^2 + (\lambda/x_0)^2]^2} \qquad [2.10.1]$$

and

$$G''(\omega) = \frac{4\pi\,\omega^2\,\lambda^2\,\rho\,(\lambda/x_0)}{[4\pi^2 + (\lambda/x_0)^2]^2} \qquad [2.10.2]$$

where ρ is the density of the medium, λ is the wavelength of the visco-elastic wave and x_0 is the distance from the plate for the wave to be damped to $1/e$ of its initial value.

The frequency range is from 10^2 to 10^5 c/s.

(2) *Characteristic impedance measurements*

When an elastic wave is propagated into a fluid sample so large that the amplitude of the wave has decayed to zero before it reaches the boundary of the containing vessel (no reflection) the quantities G' and G'' can be calculated from the complex ratio of force to velocity at the surface from which the wave is generated.

Mason[27] has described such an instrument. In this a hollow cylindrical piezoelectric crystal immersed in the liquid is excited in torsional oscillations. The complex ratio of force to velocity at the surface can be deduced from the electrical resistance and reactance of the crystal. The frequency range is 10^3 to 10^7 c/s.

2.11 ANALYSIS OF EXPERIMENTAL DATA FOR VISCOELASTIC SYSTEMS

In the preceding section various methods have been described which are designed to specify the properties of viscoelastic materials. It has been found convenient to correlate the results of both transient and dynamic experiments in terms of the distribution functions of relaxation times (for a liquid) or retardation times (for a solid). Therefore by analysing the results of experiments we try to characterize the material by a single parameter, either $J(\lambda)$ for a solid or $G(\lambda)$ for a liquid.

Although the prediction of experimental results from a knowledge of either $J(\lambda)$ or $G(\lambda)$ is relatively easy, the reverse procedure, i.e. the deduction of $J(\lambda)$ or $G(\lambda)$ from experimental results, is considerably more difficult. In principle these functions could be obtained by exact inversion mathematically as indicated in section 1.5(*f*), but in practice this method is impracticable in most cases because the experimental data are not usually sufficiently complete. Approximate methods are available, however.

As a general rule the distribution of relaxation times, $G(\lambda)$, which conveniently characterizes a viscoelastic liquid, is found from relaxation tests or from the real or imaginary part of $G^*(\omega)$. $J(\lambda)$ for solids is most conveniently derived from creep measurements or from the real or imaginary part of $J^*(\omega)$. The approximate expressions for $G(\lambda)$ and $J(\lambda)$ derived in this way will be summarized.

(a) Determination of G(λ)

(1) *from stress relaxation function* $\psi(t)$

Andrews[28] has shown that the distribution of relaxation times $G(\lambda)$ can be calculated from the relaxation function $\psi(t)$, obtained by experiment, from the relation

$$G(\lambda) = -\,\mathrm{d}\,\psi(t)/\mathrm{d}t \text{ at } \lambda = t \qquad [2.11.1]$$

This implies that $G(\lambda)$ at any value of λ is simply the negative slope of the relaxation curve at the same value of t. This is only a first approximation but is accurate if the distribution $G(\lambda)$ is fairly broad.

(2) *from dynamic experiments*

$G(\lambda)$ is conveniently obtained from the in-phase component of $G^*(\lambda)$, to a first approximation by the equation

$$[G(\lambda)]_{\lambda=\frac{1}{\omega}} = -\frac{\mathrm{d}\,G'(\omega)}{\mathrm{d}\,(1/\omega)} \qquad [2.11.2]$$

This means that if $G'(\omega)$ is plotted against $1/\omega$ (where $1/\omega$ is the *circular* frequency) the value of G at any time $1/\omega$ is simply the slope of this curve at this value of $1/\omega$.

(b) Determination of J(λ)

(1) *from creep experiments*

In a similar fashion to the evaluation of $G(\lambda)$ from stress relaxation curves it can be shown that $J(\lambda)$ can be derived to a first approximation from the creep curve using the expression

$$[J(\lambda)]_{\lambda=t} = \frac{\mathrm{d}\,\phi(t)}{\mathrm{d}t} - \frac{1}{\mu} \qquad [2.11.3]$$

where $\phi(t)$ is the creep function and μ is the steady flow viscosity (if finite).

(2) *from dynamic experiments*

$J(\lambda)$ is conveniently obtained as a first approximation from the in-phase component of $J^*(\omega)$ according to the relation

$$[J(\lambda)]_{\lambda=\frac{1}{\omega}} = -\frac{\mathrm{d}\,J'(\omega)}{\mathrm{d}\,(1/\omega)} \qquad [2.11.4]$$

i.e. by taking slopes from a graph of $J'(\omega)$ versus $1/\omega$.

(c) Higher order approximations for G(λ) and J(λ)

The expressions [2.11.1] to [2.11.4] are merely first approximations. For distributions which are broad these are usually adequate, but if the distributions show a sharp peak, approximations of a higher order have to be used. These have been fully discussed by Andrews [28] and also by Schwartzl and Staverman. [29]

2.12 EXPERIMENTAL CHARACTERIZATION OF VISCOELASTIC MATERIALS BY MEASUREMENT OF NORMAL STRESS

Consider an element of fluid in pure shear as shown in Fig. 30. The normal stresses on the three planes are p_{11}, p_{22}, and p_{33} and the shearing stress in the plane of shear is τ_{21}. The complementary shear stress is τ_{12}.

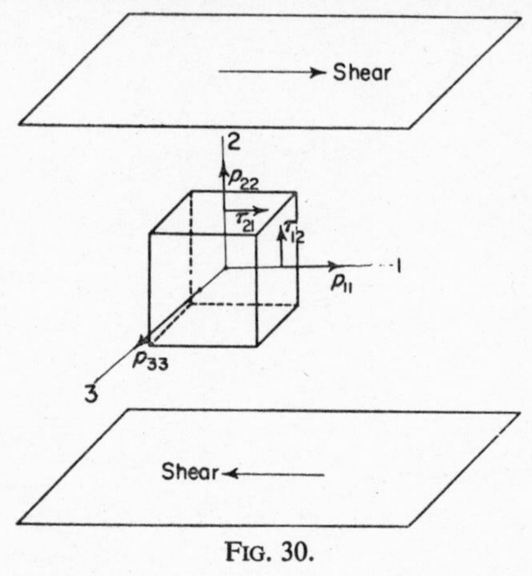

FIG. 30.

If the fluid is viscoelastic a normal stress, shown as p_{22}, will be developed. This stress is perpendicular to the streamlines in the plane of shear and normal to the plane of shear.

Roberts [30] has developed an experimental method which is capable of characterizing viscoelastic liquids (and also distinguishing them from inelastic liquids) and which is based on this principle. The material is sheared between a rotating cone and a stationary plate to ensure a uniform rate of shear throughout the specimen, and the normal stresses, p_{22} which are characteristic of viscoelastic liquids, are measured by balancing the normal pressures developed against a head of liquid in capillary tubes inserted in the stationary plate. The stresses p_{33} were also measured. The apparatus is shown diagrammatically in Fig. 31 and it is described in more detail in Chapter 6.

Roberts found that in all the systems he investigated

$$p_{22} = p_{33} \qquad [2.12.1]$$

Using the principle of the equilibrium of forces on the element it may be shown that

$$p_{11} = \mathrm{d}\, p_{33}/\mathrm{d} \ln (r/R) + 2p_{33} - p_{22} \qquad [2.12.2]$$

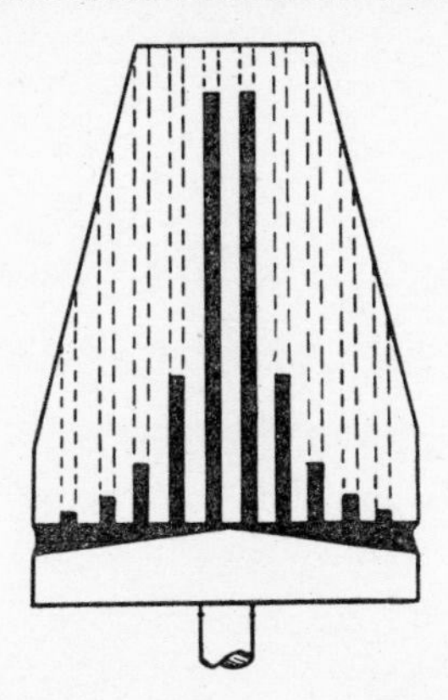

FIG. 31. Normal stress in a viscoelastic liquid.

where r is the radius under consideration and R is the radius of the free boundary.

From these equations we get

$$p_{11} - p_{22} = p_{11} - p_{33} = \mathrm{d}\, p_{33}/\mathrm{d} \ln (r/R) \qquad [2.12.3]$$

These experimental findings are consistent with the theory of Weissenberg[31] which may be written

$$p_{11} - p_{22} = p_{11} - p_{33} = \mathrm{d}\, p_{33}/\mathrm{d} \ln (r/R) = \tau_{21}\gamma_e \qquad [2.12.4]$$

where γ_e is the shearing component of recoverable strain.

The normal pressure effect or 'Weissenberg effect' is associated with the tension along the streamlines caused by the shearing flow if the material is viscoelastic. If the geometry of flow has cylindrical symmetry this tension along the streamlines becomes a hoop stress. This causes the fluid to be constricted towards the cylindrical axis and the effect is well demonstrated by the fact that a viscoelastic liquid will climb up a rod which is rotated in it.

CHAPTER 3

FLOW OF NON-NEWTONIAN FLUIDS IN PIPES AND CHANNELS

In this chapter the flow of time-independent non-Newtonian fluids in circular pipes will be treated in detail. The case of laminar flow will be dealt with first, since the analysis in this case is relatively simple, and then formulae will be developed which allow the prediction of friction factors and velocity profiles for turbulent flow in smooth pipes. The analogous case of turbulent flow in rough pipes will also be discussed briefly. The laminar flow of time-independent fluids in non-circular ducts, such as an annulus, or an extruder screw, and passage between rolls, will also be treated.

Although time-dependent fluids, particularly of the thixotropic type, are of considerable commercial importance, virtually no work of an engineering nature has appeared in the literature and a section on these fluids in the present chapter would merely serve to emphasize the lack of information in this field. This is not unduly surprising when one considers that processing equipment has to be designed to operate at the extremes of the fluid physical properties. For example if a thixotropic fluid has been left to stand in a pipe the pump must be sufficiently powerful to start flow. Once flow has started the material will break down under shear and the load on the pump will be considerably reduced. After the fluid has been sheared for a long time it will become time-independent. Hence at the limiting conditions encountered at start-up and in the final steady state a thixotropic fluid is not materially different from the simple time-independent fluid. The same remarks would presumably apply to rheopectic materials but in this case in the above example the critical load on the pump would be developed slowly and not at start-up. It is doubtful, however, if many truly rheopectic fluids will be encountered in industry.

Likewise in *steady* laminar flow viscoelastic fluids are unlikely to present special problems. However, when unsteady state conditions exist, such as in the flow through valves or fittings and in the whole field of turbulent flow, the elastic properties of the fluid would be of major importance.

3.1 RELATIONS BETWEEN THROUGHPUT AND PRESSURE DROP FOR LAMINAR FLOW IN CIRCULAR PIPES

If the fluid properties are independent of time the rheological equation relating shear stress and shear rate may be written

$$\dot{\gamma} = f(\tau)$$

For flow in a pipe this becomes

$$-\,\mathrm{d}u/\mathrm{d}r = f(\tau) \qquad [3.1.1]$$

where τ is now the shear stress at radius r.

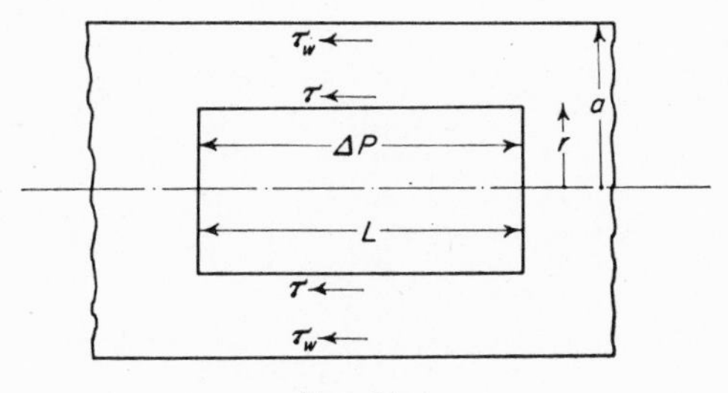

FIG. 32.

The balance of forces on a cylindrical element of fluid of radius r and length L (Fig. 32) gives

$$2\pi r\,L\,\tau = \pi r^2 \Delta P$$

or

$$\tau = \frac{r\Delta P}{2L}$$

and for the shear stress at the wall, τ_ω, we have

$$\tau_\omega = \frac{a\Delta P}{2L}$$

Therefore $\tau = \tau_\omega r/a$ and Eqn. [3.1.1] becomes

$$-\,\mathrm{d}u/\mathrm{d}r = f(\tau_\omega r/a)$$

Integrating we get

$$u(r) = \int_r^a f(\tau_\omega r/a)\,\mathrm{d}r \qquad [3.1.2]$$

since $u(a) = 0$ if we assume no slip at the walls.

Further we have

$$Q = \int_0^a 2\pi\, r\, u(r)\, \mathrm{d}r$$

or

$$Q = \pi \int_0^a u(r)\, \mathrm{d}(r^2)$$

Integrating by parts we get

$$Q = \pi \left[r^2\, u(r) - \int r^2\, \mathrm{d}\, u(r) \right]_0^a$$

i.e.

$$Q = \pi \int_0^a r^2\, f(\tau_\omega r/a)\, \mathrm{d}r \text{ since } u(a) = 0$$

Substituting $r = a\tau/\tau_\omega$ we get

$$\frac{Q}{\pi a^3} = \frac{1}{\tau_\omega^3} \int_0^{\tau_\omega} \tau^2 f(\tau)\, \mathrm{d}\tau \qquad [3.1.3]$$

The relations between flow and pressure drop may be derived from Eqn. [3.1.3] by numerical integration from the experimentally determined form of $f(\tau)$. When $f(\tau)$ is a simple expression the integration can be performed analytically as follows.

(a) Newtonian fluid

For a Newtonian fluid in laminar flow we have

$$\tau = \mu\, \dot{\gamma}$$

i.e.

$$f(\tau) = \tau/\mu$$

Substituting in Eqn. [3.1.3] we get

$$\frac{Q}{\pi a^3} = \frac{1}{\mu \tau_\omega^3} \int_0^{\tau_\omega} \tau^3\, \mathrm{d}\, \tau$$

which on integration gives

$$Q = \pi a^3\, \tau_\omega / 4\mu$$

Substituting $\tau_\omega = a\Delta P/2L$ we get the familiar Poiseuille equation for Newtonian laminar flow

$$Q = \pi a^4 \Delta P/8\mu L \qquad [3.1.4]$$

(b) Bingham plastic

For a Bingham plastic we have

$$\dot{\gamma} = (\tau - \tau_y)/\mu_p = f(\tau) \; ; \tau > \tau_y$$

$f(\tau)$ is discontinuous and

$$f(\tau) = 0 \; ; 0 < \tau < \tau_y$$

and
$$f(\tau) = (\tau - \tau_y)/\mu_p \; ; \tau_y < \tau < \tau_\omega$$

In pipe flow the shearing stress falls to zero at the axis, and in a region near the axis, where the local shearing stress is less than the yield value, τ_y, the material does not shear but moves along as a solid plug. This is illustrated in Fig. 33.

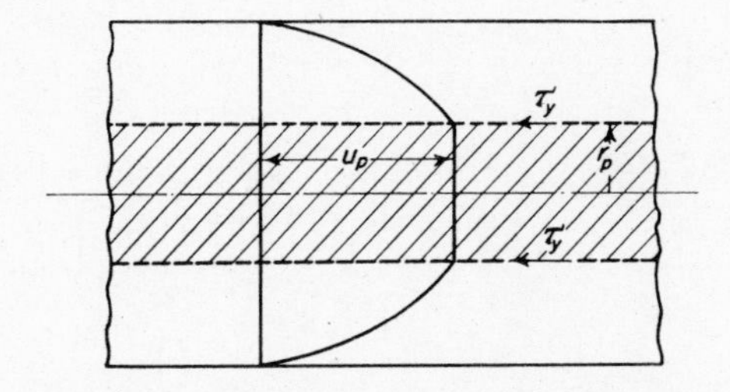

FIG. 33. Velocity profile for a Bingham plastic.

Substituting for $f(\tau)$ in Eqn. [3.1.3.] we get

$$\frac{Q}{\pi a^3} = \frac{1}{\mu_p \tau_\omega^3} \int_{\tau_y}^{\tau_\omega} \tau^2(\tau - \tau_y) \, d\tau$$

which on integration gives

$$\frac{Q}{\pi a^3} = \frac{1}{\mu_p \tau_\omega^3} \left[\frac{\tau^4}{4} - \frac{\tau^3 \tau_y}{3} \right]_{\tau_y}^{\tau_\omega}$$

and after inserting the limits we get

$$\frac{Q}{\pi a^3} = \frac{\tau_\omega}{\mu_p}\left[\frac{1}{4} - \frac{1}{3}\left(\frac{\tau_y}{\tau_\omega}\right) + \frac{1}{12}\left(\frac{\tau_y}{\tau_\omega}\right)^4\right] \qquad [3.1.5]$$

Substituting for τ_ω gives

$$Q = \frac{\pi a^4 \Delta P}{8L\mu_p}\left[1 - \frac{4}{3}\left(\frac{2L\tau_y}{a\Delta P}\right) + \frac{1}{3}\left(\frac{2L\tau_y}{a\Delta P}\right)^4\right] \qquad [3.1.6]$$

This is known as Buckingham's equation and it is seen that it cannot be solved explicitly for the pressure loss. It reduces to the Poiseuille equation when the yield stress is zero. Caldwell and Babbitt (32) have applied the Buckingham equation successfully to the flow of muds and sludges.

McMillen (33) simplified the computations for Bingham plastic flow by rearranging the Buckingham equation in dimensionless form, the groups having numerical values which depend only on the relative diameters of the central unsheared plug (in which the shear stress is less than the yield value) and the layer of fluid under consideration. From a knowledge of the yield value and the plastic viscosity (which may be obtained in principle by two determinations of pressure drop at two different rates of flow in a pipe of any diameter) the relation between pressure loss and throughput may be found for laminar flow in any size of pipe.

Hedström (34) has produced a very convenient graphical equivalent of the Buckingham equation for obtaining the pressure drop. This method is based on dimensional analysis. If we assume for a Bingham plastic that

$$\Delta P = \phi(\rho, \mu_p, \tau_y, D, u_m, L)$$

dimensional analysis will give a functional relation of the form

$$c_f = \frac{D\Delta P/4L}{\rho u_m^2/2} = \phi\left[\frac{\rho u_m D}{\mu_p}, \frac{\tau_y \rho D^2}{\mu_p^2}\right] \qquad [3.1.7]$$

which states that the friction factor is a function of a Reynolds number $\rho u_m D/\mu_p$ and the group $\tau_y \rho D^2/\mu_p^2$, normally referred to as the Hedström group, He. The correlation, shown in Fig. 34, consists of a plot of c_f versus $\rho u_m D/\mu_p$ giving a family of curves in the laminar region for various values of the Hedström number. Alternatively, the plasticity number $\tau_y D/\mu_p u_m$ could be used as the parameter.

Ooyama and Ito (35) have also presented a design method for Bingham plastic fluids but this involves a trial and error solution and is not so convenient as the method of Hedström. (34)

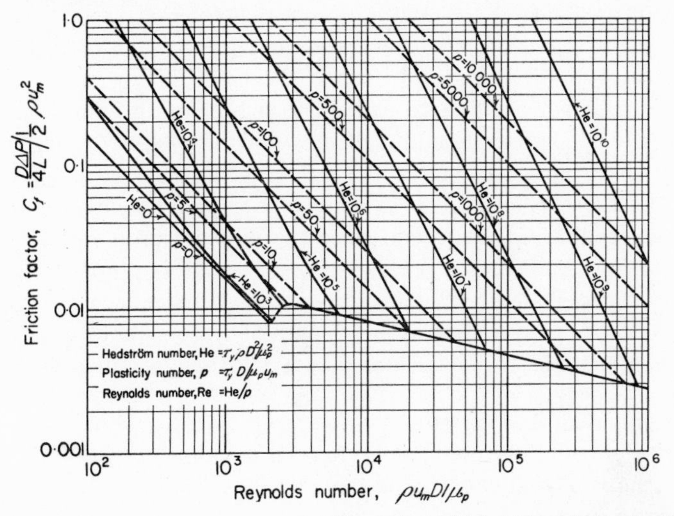

(By courtesey of *Industrial and Engineering Chemistry*.)

FIG. 34. Friction factor v. Reynolds number diagram for Bingham plastics.

(c) Power law fluids

For a power law fluid we have

$$\tau = k\dot{\gamma}^n$$

Therefore

$$f(\tau) = (\tau/k)^{1/n}$$

Hence

$$\frac{Q}{\pi a^3} = \frac{1}{k^{1/n}\tau_\omega^3}\int_0^{\tau_\omega} \tau^{2+1/n}\,\mathrm{d}\,\tau$$

and after integrating and substituting for τ_ω we get

$$Q = \frac{n\ \pi\ a^3}{3n+1}\left(\frac{a\Delta P}{2Lk}\right)^{1/n} \qquad [3.1.8]$$

For a Newtonian fluid $n = 1$, $k = \mu$ and the equation again reduces to the Poiseuille form.

An interesting result for pseudoplastic fluids ($n < 1$) may be illustrated by writing Eqn. [3.1.8] in the form

$$\Delta P \propto LQ^n/a^{3n+1}$$

For Newtonian fluids $n = 1$ and it is seen that for a given flow rate ΔP is proportional to $1/a^4$, i.e. small increases in pipe size are very effective in reducing pressure drop. On the other hand for highly non-Newtonian pseudoplastics n is close to zero and ΔP then becomes proportional to $1/a$, rather than $1/a^4$. However, although it would now be necessary to go to unusually large pipe sizes before the reduction in pressure drop becomes appreciable, the capacity of an existing pipe may frequently be increased merely by speeding up the pump, since the pressure drop will be very insensitive to flow rate when n is close to zero, for ΔP is proportional to Q^n. The same argument indicates that flow-metering devices which involve the measurement of pressure drop over a length of pipe would be unsatisfactory for highly pseudoplastic fluids since they would be insensitive to flow rate.

(d) Other empirical flow curves

The principle of all the methods (*a*) to (*e*) is to substitute an empirical form of the function $f(\tau)$ that closely approximates the experimental flow curve and is sufficiently simple in form to be capable of mathematical manipulation. The power law is the most widely applicable relation for non-Newtonian fluids but others have been suggested.

Williamson[36] gave an empirical expression relating shear stress and shear rate in the form

$$\tau = \frac{A\dot{\gamma}}{B+\dot{\gamma}} + \mu_{\infty}\dot{\gamma} \qquad [3.1.9]$$

In this equation, A is the extrapolated intercept of the linear part of the curve at high shear rates on the stress exis, equivalent to the Bingham yield value, and B is a constant which is a measure of the curvature of the flow curve. Campbell[37] has applied this equation to the flow of molten chocolate.

Powell and Eyring[38] suggested the empirical equation

$$\tau = A\dot{\gamma} + B \sinh^{-1}(C\dot{\gamma}) \qquad [3.1.10]$$

where the constants A, B and C are characteristic of the fluid. Application of dimensional analysis to the pipe flow problem with Q, $\Delta P/L$, a and the fluid properties A, B and C as variables predicts the functional relation

$$\frac{QC}{a^3} = \phi\left[\frac{a\Delta P}{2LB}, \frac{A}{BC}\right] \qquad [3.1.11]$$

Christiansen *et al.*[39] derived this relation numerically, and from the results prepared a graph of QC/a^3 plotted against $a\,\Delta P/2LB$ for various values of

A/BC. This chart enables one of the three variables a, $\Delta P/L$ and Q to be calculated from a knowledge of the other two. The constants A, B and C in the rheological equation [3.1.10] may be found from either rotational or tube-viscometer data for the fluid under test. The method was applied successfully to suspensions of lime and solutions of carboxy-methyl cellulose in water.

Many other attempts have been made to fit an empirical equation to the flow curve. Most of these have been summarized by Scott-Blair[(40)] The objection to all these methods, and others of a similar nature, is that empirical equations of a conveniently simple form seldom represent the data over the full range of shear rate with the desired accuracy. It cannot be stressed too strongly that large extrapolations of data are likely to lead to erroneous conclusions because the rheological properties often depend upon the experimental conditions under which the data were obtained. A fluid which behaves as a Bingham plastic in one shear range could easily become pseudo-plastic in another, and at high rates of shear appear Newtonian. For this reason methods involving the classification of fluids into types are not generally satisfactory.

(e) General methods applicable to all fluids

It becomes apparent from the above discussion that engineering design procedures would be greatly simplified if some method could be developed which would be universally applicable to all fluids in the laminar region, whether they are Newtonian or non-Newtonian.

The first attempt in this direction was made by Alves, Boucher and Pigford.[(41)] Their method depends on the fact previously discussed in Section 2.3 of Chapter 2 that, for fluids with rheological properties independent of time, the relation between $Q/\pi a^3$ (or $8\ u_m/D$) and $a\Delta P/2L$ is unique and independent of pipe size. Once this relation has been determined over the range of $Q/\pi a^3$ of interest, e.g. in a capillary tube viscometer, the scale-up to plant size at a constant value of $Q/\pi a^3$ can be made directly. If possible, measurements using at least two capillary tubes of different diameters should be made in order to ensure the absence of time-dependent properties and anomalous flow behaviour near the wall. If either of these complications exist the plot of $Q/\pi a^3$ versus $a\Delta P/2L$ will depend on tube size.

Metzner[(42)] has suggested a similar method of scaling-up for the correlation and prediction of pressure drops in pipes for time-independent non-Newtonian fluids. The usefulness of this method is restricted by its empirical nature but it has three main applications as follows.

(1) It may be used as a basis for the design of a model with which to predict the pressure-drop characteristics of the prototype. The criteria for the

design of the model are that it must operate with the same ratio of fluid velocity to pipe diameter and with the same fluid as the prototype.

(2) Laboratory data obtained with ordinary non-ideal viscometers may be used directly to predict the effect in changes of fluid properties or pressure drop in a pipe.

(3) Since the correlation is in terms of dimensionless groups it allows limited plant data to be extended and generalized.

A more general method has since been presented by Metzner and Reed.(19) This method springs from Mooney's expression for the shear stress at the pipe wall (Eqn. [2.4.1]) which may be rearranged to give

$$-\left(\frac{\mathrm{d}u}{\mathrm{d}r}\right)_\omega = \frac{3n' + 1}{4n'}\frac{8u_m}{D} \qquad [2.4.3]$$

where

$$n' = \frac{\mathrm{d}\,(\ln D\Delta P/4\,L)}{\mathrm{d}\,(\ln 8\,u_m/D)} \qquad [2.4.4]$$

Further we can write

$$\frac{D\Delta P}{4L} = k'\left(\frac{8\,u_m}{D}\right)^{n'} \qquad [2.4.5]$$

It is important to note that Eqn. [2.4.5] is not an integration of Eqn. [2.4.4], but merely a definition of k', the consistency index. Only if n' is constant can Eqn. [2.4.5] be obtained by direct integration of Eqn. [2.4.4].

The friction factor, c_f, is defined in the usual manner by

$$c_f = \frac{D\Delta P}{4L}\bigg/\frac{\rho u_m^2}{2} \qquad [3.1.12]$$

and in order to be able to use the conventional chart for Newtonian and non-Newtonian fluids alike in the laminar region it is convenient to let $c_f = 16/\mathrm{Re}'$, as for Newtonians in laminar flow. The generalized Reynolds number is then found by combining Eqns. [3.1.12] and [2.4.5] to give

$$\mathrm{Re}' = \mathrm{D}^{n'}\,u_m{}^{2-n'}\,\rho/k'\,8^{n'-1}$$

This quantity is dimensionless.

For a power law fluid we can put the Reynolds number in terms of n and k rather than n' and k', since then we have

$$n' = n$$

$$\text{and } k' = k(3n + 1/4n)^n$$

Hence substituting we get

$$\text{Re}' = \frac{D^n u_m^{2-n} \rho}{\frac{k}{8}\left(\frac{6n+2}{n}\right)^n}$$

In this way data for all time-independent fluids should follow the conventional $c_f = 16/\text{Re}'$ relation rigorously in the laminar region if the fluid flow parameters n' and k' are evaluated at the actual value of $8\ u_m/D$ under consideration. This is shown in Fig. 35.

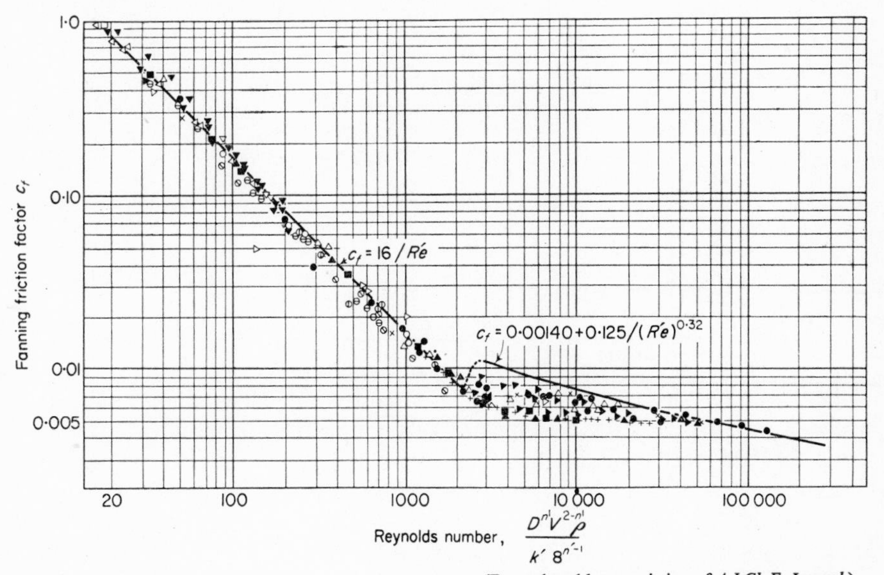

(Reproduced by permission of *A.I.Ch.E. Journal.*)

FIG. 35. Friction factor v. Reynolds number diagram for non-Newtonian fluids in the laminar region.

The advantage of this method over the use of Eqn. [3.1.8] is that the latter is derived by integration of the power law with the implication that the fluid has a constant value of the flow index, n, over the whole range of shear rates encountered in the pipe. The method of Metzner and Reed involves no such assumption and is theoretically rigorous for all time-independent fluids (including Bingham plastics).

Weltmann[43] has recently suggested a similar correlation for non-Newtonian fluids on the conventional friction-factor diagram. In this case the Reynolds number was defined as

$$\text{Re} = \frac{D u_m \rho}{\text{'viscosity'}}$$

Rheological Properties of Fluids Shown in Fig. 35

Symbol	*Nominal pipe size (in.)*	*Composition of fluid*	*Rheological properties*	
			n'	$m = k' \, 8^{n'-1}$
+	1	23·3% Illinois yellow clay in water	0·229	0·863
⦶	$\frac{7}{8}$ and $1\frac{1}{2}$	0·67% carboxy-methyl-cellulose (CMC) in water	0·716	0·121
⊖	$\frac{7}{8}$ and $1\frac{1}{2}$	1·5% CMC in water	0·554	0·920
⊘	$\frac{7}{8}$ and $1\frac{1}{2}$	3·0% CMC in water	0·566	2·80
⦸	$\frac{7}{8}$, $1\frac{1}{2}$ and 2	33% lime water	0·171	0·983
◁	$\frac{7}{8}$ and $1\frac{1}{2}$	10% napalm in kerosene	0·520	1·18
▼	8, 10 and 12	4% paper pulp in water	0·575	6·13
△	$\frac{3}{4}$ and $1\frac{1}{2}$	54·3% cement rock in water	0·153	0·331
▲	4	18·6% solids, Mississippi clay in water	0·022	0·105
●	$\frac{3}{4}$ and $1\frac{1}{4}$	14·3% clay in water	0·350	0·0344
▷	$\frac{3}{4}$ and $1\frac{1}{4}$	21·2% clay in water	0·335	0·0855
×	$\frac{3}{4}$ and $1\frac{1}{4}$	25·0% clay in water	0·185	0·204
▽	$\frac{3}{4}$ and $1\frac{1}{4}$	31·9% clay in water	0·251	0·414
□	$\frac{3}{4}$ and $1\frac{1}{4}$	36·8% clay in water	0·176	1·07
■	$\frac{3}{4}$ and $1\frac{1}{4}$	40·4% clay in water	0·132	2·30
▶	$\frac{1}{8}$, $\frac{1}{4}$, $\frac{1}{2}$ and 2	23% lime in water	0·178	1·04

the 'viscosity' being the Newtonian viscosity for Newtonian fluids, the plastic viscosity, μ_p, for a Bingham plastic, and the apparent viscosity, μ_a, for a pseudoplastic or dilatant fluid, measured at the prevailing flow condition in the pipeline. On this basis the friction factor is related to the Reynolds number in the laminar region for a Newtonian fluid by 16/Re. For a Bingham plastic the friction factor is given by

$$c_f = \frac{16}{\text{Re}} \frac{p}{8c}$$

where c is the ratio of the yield value to the shearing stress at the pipe wall and $p = \tau_y D / \mu_p u_m$ is called the *plasticity number*, as it determines the degree of plastic behaviour. For pseudoplastic and dilatant materials for which $\tau \propto (\mathrm{d}u/\mathrm{d}r)^n$ the friction factor is given by

$$c_f = \frac{16}{\text{Re}} \left(\frac{3n + 1}{4n} \right)$$

c_f is then plotted over a range of Reynolds number for various values of p and n as shown in Fig. 36.

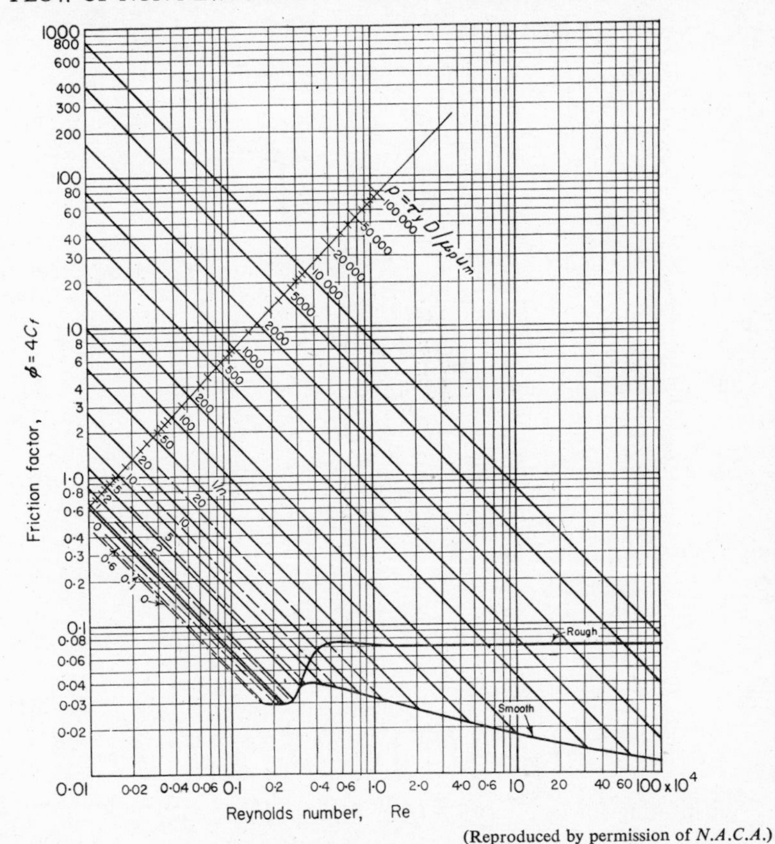

(Reproduced by permission of *N.A.C.A.*)

FIG. 36. Friction faction v. Reynolds number diagram for Bingham plastics and pseudo-plastics in laminar region.

This approach by Weltmann is based on the concept of an ideal fluid, either a Bingham plastic or a true power law fluid, and as a result is not as general as the method of Metzner and Reed. In fact for pseudoplastic fluids the Weltmann correlation is virtually a special case of the Metzner and Reed correlation when n' is constant. It also has the disadvantage that it consists of a family of separate curves with the plasticity number or flow behaviour index as parameter. The Metzner and Reed method correlates the friction factors for all time-independent fluids in the laminar region by a single straight line.

3.2 VELOCITY PROFILES IN LAMINAR FLOW

Consider the balance of forces on a cylindrical element of fluid (Fig. 32):

$$\pi r^2 \Delta P = 2\pi rL\tau$$

or

$$\tau = r\Delta P/2L \qquad [3.2.1]$$

For a Newtonian fluid we have

$$\tau = -\mu \mathrm{d}u/\mathrm{d}r \qquad [3.2.2]$$

therefore

$$-\int_u^0 \mathrm{d}u = \frac{\Delta P}{2\mu L}\int_r^a r\,\mathrm{d}r$$

if $u = 0$ at $r = a$, and the parabolic velocity distribution is given on integration by

$$u = \frac{\Delta P}{4\mu L}(a^2 - r^2)$$

Substituting from Eqn. [3.1.4] for $u_m = Q/\pi a^2$ we get

$$u = 2u_m(1 - r^2/a^2) \qquad [3.2.3]$$

For a power law fluid we have in place of Eqn. [3.2.2]

$$\tau = -k(\mathrm{d}u/\mathrm{d}r)^n$$

giving from Eqn. [3.2.1]

$$u = \left(\frac{\Delta P}{2Lk}\right)^{1/n}\int_r^a r^{1/n}\,\mathrm{d}r$$

i.e.

$$u = \left(\frac{n}{n+1}\right)\left(\frac{\Delta P}{2Lk}\right)^{1/n}\left(a^{(1/n)+1} - r^{(1/n)+1}\right)$$

Substituting for $u_m = Q/\pi a^2$ from Eqn. [3.1.8] we get

$$u = u_m\left(\frac{3n+1}{n+1}\right)\left[1 - \left(\frac{r}{a}\right)^{(n+1)/n}\right] \qquad [3.2.4]$$

from which we can derive Eqn. [3.2.3] by putting $n = 1$.

Typical velocity profiles based on Eqn. [3.2.4] are drawn in Fig. 37 for a Newtonian fluid, $n = 1$, a dilatant fluid, $n = 3$, and a pseudoplastic fluid with $n = 1/3$. Profiles for infinite pseudoplasticity and infinite dilatancy, represented by $n = 0$ and ∞ respectively, are also plotted.

For a Bingham plastic we have in place of Eqn. [3.2.2]

$$\tau - \tau_y = -\mu_p(\mathrm{d}u/\mathrm{d}r)$$

i.e.

$$r\Delta P/2L - \tau_y = -\mu_p(\mathrm{d}u/\mathrm{d}r)$$

and on integration this gives

$$u = \frac{1}{\mu_p}\left[\frac{(a^2 - r^2)\Delta P}{4L} - \tau_y(a - r)\right] \qquad [3.2.5]$$

if $u = 0$ at $r = a$.

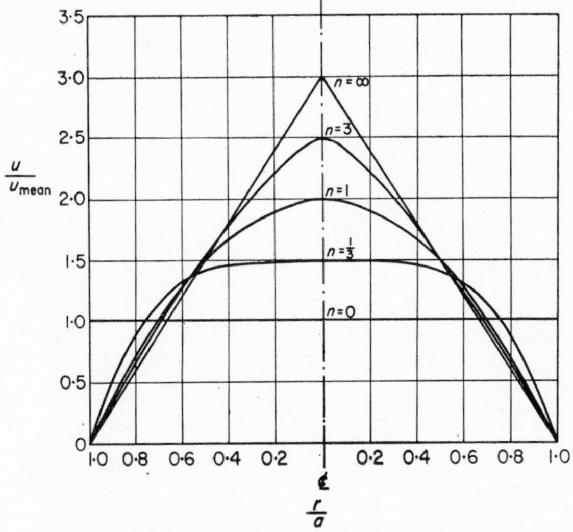

FIG. 37. Velocity profiles for non-Newtonian fluids.

Near the axis where $\tau < \tau_y$ the material will move as a solid cylindrical plug and the radius of the plug will be given by

$$r_p = 2L\tau_y/\Delta P$$

Substituting this value of r_p into Eqn. [3.2.5] we get the velocity of the plug as

$$u_p = \frac{\Delta P}{2L\mu_p}(a - r_p)^2 \qquad [3.2.6]$$

For given values of $\Delta P/L$, μ_p and τ_y the velocity profile could be calculated from Eqns. [3.2.5] and [3.2.6]. The corresponding flow-rate would be given by Eqn. [3.1.6].

3.3 TURBULENT FLOW OF TIME-INDEPENDENT FLUIDS IN CIRCULAR PIPES

(a) Review of early work

Most of the early studies which attempted to correlate the relation between the turbulent friction factor and flow rate for non-Newtonian systems have been largely empirical. In many cases the work has been restricted to particular types of fluid behaviour, such as Bingham plastic, and this has led to a variety of suggested design procedures of limited applicability.

It has usually been pointed out that the apparent viscosity of many non-Newtonian fluids under turbulent conditions is nearly constant and, as a result, most of the early workers have suggested that the conventional diagram relating friction factor and Reynolds number, which has been developed for

Newtonian systems, is also applicable to non-Newtonian systems if a realistic value is chosen for the viscosity term in the Reynolds number. The basis for this approach is the well-known fact that non-Newtonian fluids of the pseudoplastic and Bingham plastic types approach a limiting viscosity at sufficiently high rates of shear and thereafter appear Newtonian. The implication in the early work is then that the shear rates occurring in the turbulent flow region are sufficiently high to suppress all non-Newtonian behaviour and cause the system to behave simply as Newtonian.

It is clear then that these studies have not been concerned with fluids which exhibit pronounced non-Newtonian properties at the shear rates prevailing under the turbulent flow conditions studied, but rather with the turbulent Newtonian behaviour of fluids which at lower shear rates show marked deviations from Newtonian properties.

The principal difference among these studies has been concerned with the choice of viscosity to use in the Reynolds number. Caldwell and Babbitt (32) in their work on sewage sludges and slurries used the viscosity of the dispersion medium. The basis of this assumption is the fact that with increasing shear rate the apparent viscosity of dispersions decreases and approaches that of the dispersion medium. The satisfactory correlation obtained by these workers suggests that the shear rates existing in the turbulent state were high enough to allow the limiting viscosity to be reached. It is unlikely of course that this limiting viscosity was as low as that of the dispersion medium.

Other workers, notably Alves, Boucher and Pigford (41) and also Winding, Baumann and Kranich (44), have suggested that the limiting viscosity at infinite shear rate, μ_∞, is the more realistic. These workers found that the conventional plot of friction factor against Reynolds number for Newtonian systems correlates the results for non-Newtonian systems satisfactorily if the Reynolds number is defined as $\rho u_m D/\mu_\infty$.

Weltmann (43) has reviewed work on Bingham plastics in which turbulent friction-factor data are correlated on the conventional Newtonian chart by using a Reynolds number defined by $\rho u_m D/\mu_p$. This is essentially the same as the previous approach since μ_∞ for Bingham plastics is the same thing as μ_p. However these conclusions are at variance with those of Caldwell and Babbitt (32) who suggested the use of the viscosity of the dispersion medium in the Reynolds number.

Another approach which has advantages over the method using μ_∞ involves the use of a 'turbulent viscosity'. This is found from the relation between pressure drop and flow rate for the particular fluid under turbulent conditions. The experimentally determined friction factor so obtained is used to calculate the Reynolds number from the Newtonian friction factor chart. This Reynolds number is equated to $\rho u_m D/\mu_t$ to give μ_t, the 'turbulent viscosity'. This is then assumed constant. It is clear that the disadvantage of this approach is that it does not allow the design of a pipeline directly from a knowledge of fluid properties. It is in fact a scale-up procedure whereby full-scale equipment can be designed once the empirical 'turbulent viscosity' has been determined for the given fluid in a similar pilot plant. Alves *et al.*[41] have adopted this approach.

The work of Metzner and Reed[19] represented a considerable advance over the earlier work in that they did not assume a constant viscosity. Instead they used the conventional Fanning friction factor in conjunction with the generalized Reynolds number Re′, previously defined as

$$\mathrm{Re}' = \mathrm{D}^{n'} u_m{}^{2-n'} \rho / k' 8^{n'-1}$$

The Reynolds number was so defined in order to preserve the standard laminar Newtonian relation in the laminar region, i.e.

$$c_f = \frac{16}{\mathrm{Re}'}$$

This method is strictly rigorous in the laminar region and Metzner and Reed suggested an approximate correlation based on the same principle for the turbulent region (see Fig. 35).

(b) Theoretical correlation of turbulent friction factors for smooth pipes

The drawback of all the methods discussed above (except the approach of Metzner and Reed) is that they do not take into account the possibility of non-Newtonian behaviour prevailing under the conditions of shear existing in the turbulent state.

Consider the power law fluid and assume that the pressure drop ΔP will depend on the dimensions of the pipe L and D, the mean velocity u_m and the properties of the fluid ρ, k and n.

Therefore we can write

$$\Delta P = \phi(L, D, u_m, \rho, k, n)$$

Applying the normal methods of dimensional analysis we should get

$$\frac{D\Delta P/4L}{\rho u_m^2/2} = \phi\left[\frac{D^n u_m{}^{2-n}\rho}{k}, n\right]$$

or

$$c_f = \phi[\mathrm{Re}'', n] \qquad [3.3.1]$$

since the first term in the square bracket is a form of the Reynolds number. Eqn. [3.3.1] shows that c_f is a function of n.

Dodge and Metzner[45] have recently calculated the form of this functional relation and this work is the first serious attempt to put the study of turbulence in non-Newtonian systems on a firm theoretical foundation. This work is an extension of the logarithmic resistance formula of von Karman for Newtonian systems.

This equation, which has some theoretical foundation, is usually written as

$$1/\sqrt{(c_f)} = 4\cdot0 \log_{10} (\mathrm{Re}\sqrt{(c_f)}) - 0\cdot4 \qquad [3.3.2]$$

By similar reasoning Dodge and Metzner found theoretically that for fluids of the power law type the general form of Eqn. [3.3.2] is

$$1/\sqrt{(c_f)} = \mathrm{A} \log_{10}(\mathrm{Re}' \, c_f^{\,1 - n'/2}) + B \qquad [3.3.3]$$

where the coefficient A and the constant term B are now functions of n' alone and Re′ is the generalized Reynolds number, i.e.

$$\mathrm{Re}' = \mathrm{D}^{n'} u_m^{\,2-n'} \rho/k' \, 8^{n'-1}$$

Although the general form of the functions A and B can be predicted theoretically they are finally determined by experiment. To find A, $1/\sqrt{c_f}$ is plotted against $\log_{10} (\mathrm{Re}' \, c_f^{\,1 - n'/2})$. The slope of this plot is seen from Eqn. [3.3.3] to give A. This procedure was carried out for a range of pipe sizes for fluids of various flow indices, n', to give values of A as a function of n'. Theoretical considerations suggest that these results would be correlated by plotting A against n' logarithmically. When this was done it was found that the experimental data fell on a straight line of slope 0·75 and further the condition that $A = 4\cdot0$ when $n' = 1$ was satisfied in agreement with Eqn. [3.3.2].

Hence in general

$$A = 4\cdot0/(n')^{0\cdot75} \qquad [3.3.4]$$

This equation also satisfies the theoretical restrictions which are imposed on the form of A.

The function B is found by using these values of A in the equation

$$B = \frac{1}{\sqrt{c_f}} - \frac{4\cdot0}{(n')^{0\cdot75}} \log_{10} [\mathrm{Re}' \, c_f^{\,1 - n'/2}] \qquad [3.3.5]$$

and it was found that the function B calculated in this way was related to n' by the equation

$$B = -\,0\cdot4/(n')^{1\cdot2} \qquad [3.3.6]$$

Eqn. [3.3.6] satisfies the condition that $B = -\,0\cdot4$ when the fluid is Newtonian i.e. $n' = 1$.

The final correlation may now be written as

$$\frac{1}{\sqrt{(c_f)}} = \frac{4\cdot0}{(n')^{0\cdot75}} \log_{10} [\mathrm{Re}' \, c_f^{\,1 - n'/2}] - \frac{0\cdot4}{(n')^{1\cdot2}} \qquad [3.3.7]$$

which is the general form of the von Karman formula given in Eqn. [3.3.2].

The above equations were derived for power-law fluids. However it was pointed out that they are also valid for fluids which do not obey the power law provided that n' and k' are evaluated at the existing wall shear stress. This, however, is normally the quantity which is being calculated and the correct solution must be found by trial and error. This is not a serious difficulty in practice because most fluids obey the power law approximately and in these cases the convergence of the trial and error solution is rapid.

A design chart based on the results of Eqn. [3.3.7] is presented in Fig. 38b.

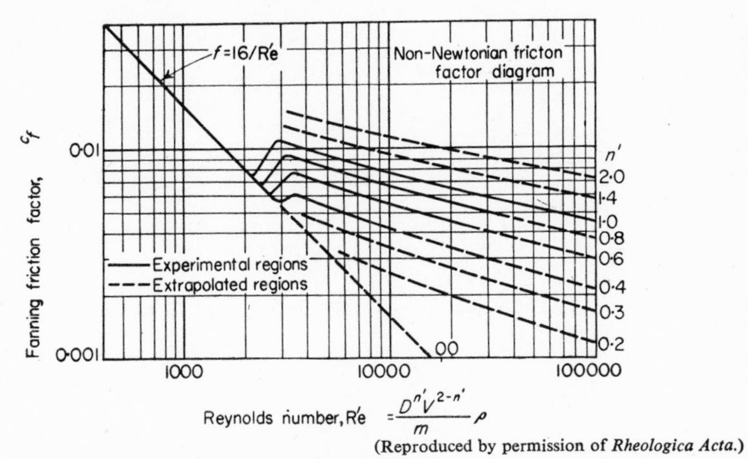

(Reproduced by permission of *Rheologica Acta.*)

FIG. 38(a) Friction factor v. Reynolds number diagram for non-Newtonian fluids in laminar and turbulent regions.

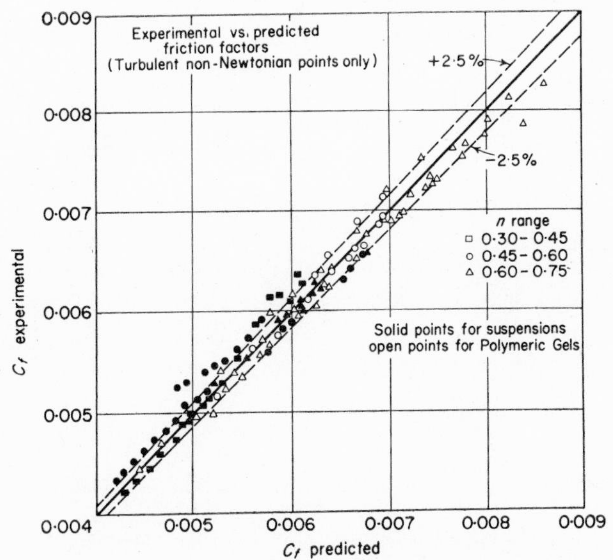

(Reproduced by permission of *Rheologica Acta.*)

FIG. 38(b). Experimental verification of Eqn. [3.3.7].

The range of experimental confirmation which has so far been achieved is also indicated on this chart.

The most significant difference between this chart and the suggested correlation of Metzner and Reed mentioned above is that there is a distinct relation in the turbulent region for each value of n'.

Eqn. [3.3.7] does not give c_f explicitly, but since the relation in the turbulent region is nearly linear for each value of n' (Fig. 38) an approximate solution of the Blasius type suggests itself. For Newtonian fluids the Blasius solution, which holds for Reynolds numbers less than 10^5, takes the form

$$c_f = 0{\cdot}079 \text{ Re}^{-0{\cdot}25} \quad [3.3.8]$$

Dodge and Metzner[45] have suggested the analogous approximation for non-Newtonian fluids

$$c_f = a(\text{Re}')^{-b} \quad [3.3.9]$$

where a and b are functions of n' alone and are given in the following Table for various values of n'.

n'	a	b
0·2	0·0646	0·349
0·3	0·0685	0·325
0·4	0·0712	0·307
0·6	0·0740	0·281
0·8	0·0761	0·263
1·0	0·0779	0·250
1·4	0·0804	0·231
2·0	0·0826	0·213

If these values of a and b are used in Eqn. [3.3.9], c_f is obtained explicitly with very little loss of accuracy.

(c) Turbulent velocity profiles in smooth pipes

It could be argued on dimensional grounds that for flow at the inner part of the turbulent region, close to the edge of the laminar sub-layer, the time-average velocity u at a distance y from the wall should be some function of τ_ω, y, ρ and μ for a Newtonian fluid. By the normal methods of dimensional analysis this functional relation reduces to

$$u/\sqrt{(\tau_\omega/\rho)} = \phi[y\sqrt{(\tau_\omega/\rho)}/v] \quad [3.3.10]$$

This equation and the velocity defect law may be combined to give

$$u/\sqrt{(\tau_\omega/\rho)} = c_1 + c_2 \log_{10} [y\sqrt{(\tau_\omega/\rho)}/v] \quad [3.3.11]$$

where c_1 and c_2 are constants to be determined by experiment.

Writing the ratio of the velocity, u, to the friction velocity $\sqrt{(\tau_\omega/\rho)}$ as u^+, i.e.

$$u^+ = u/\sqrt{(\tau_\omega/\rho)} \qquad [3.3.12]$$

and putting

$$y^+ = y\sqrt{(\tau_\omega/\rho)}/v \qquad [3.3.13]$$

we get the well-known universal velocity profile for Newtonian fluids,

$$u^+ = c_1 + c_2 \log_{10} y^+ \qquad [3.3.14]$$

Experimental measurements of turbulent velocity profiles suggest that $c_1 = 5{\cdot}5$ and $c_2 = 5{\cdot}75$, i.e.

$$u^+ = 5{\cdot}5 + 5{\cdot}75 \log_{10} y^+ \qquad [3.3.15]$$

This expression is found to hold surprisingly well over the whole of the central region of the flow.

Further, in the laminar sub-layer since τ is constant and equal to τ_ω we must have a linear velocity distribution given by

$$u = \frac{y}{v}\frac{\tau_\omega}{\rho}$$

or

$$u/\sqrt{(\tau_\omega/\rho)} = y\sqrt{(\tau_\omega/\rho)}/v$$

i.e.

$$u^+ = y^+ \qquad [3.3.16]$$

Dodge and Metzner have followed a similar argument for non-Newtonian fluids and arrived at the generalized form of the so-called universal velocity profile, given in Eqn. [3.3.15] for Newtonian fluids. Their equation takes the form

$$u^+ = c_1' + c_2' \log_{10} y^+ \qquad [3.3.17]$$

where c_1 and c_2 are now functions of n, the index in the power law, and u^+ and y^+ are given by

$$u^+ = u/\sqrt{(\tau_\omega/\rho)} \qquad [3.3.18]$$

and

$$y^+ = \frac{y^{\;n}(\tau_\omega/\rho)^{1\,-\,n/2}}{k/\rho} \qquad [3.3.19]$$

Eqn. [3.3.17] can be applied to the central region of the flow. In the laminar sub-layer, which is again assumed to be so thin that $\tau = \tau_\omega$, we have

$$\tau = \tau_\omega = k\,(-\,\mathrm{d}u/\mathrm{d}y)^n_\omega = k\,(u/y)^n$$

giving finally

$$u^+ = (y^+)^{1/n} \qquad [3.3.20]$$

of which Eqn. [3.3.16] for Newtonian liquids is a special case.

The 'constants' in Eqn. [3.3.17] have been estimated by Dodge and Metzner. Their final form of this equation is

$$u^+ = \frac{5{\cdot}66}{(n')^{0.75}} \log_{10} y^+ - \frac{0{\cdot}40}{(n')^{1{\cdot}2}}$$

$$+ \frac{2{\cdot}458}{(n')^{0{\cdot}75}} [1{\cdot}960 + 1{\cdot}255\, n' - 1{\cdot}628\, n' \log_{10} (3 + 1/n')] \quad [3.3.21]$$

When $n' = 1$, i.e. for a Newtonian fluid, Eqn. [3.3.21] reduces to

$$u^+ = 5{\cdot}66 \log_{10} y^+ + 5{\cdot}1$$

which agrees closely with the normally accepted universal velocity profile for Newtonian fluids given in Eqn. [3.3.15].

(d) The velocity profile and resistance formula for turbulent flow in rough pipes

For Newtonian flow in the turbulent core of a rough pipe it could be argued on dimensional grounds that the time-average velocity u at a distance y should be some function of ρ, ϵ, τ_ω and y, and by the normal methods of dimensional analysis we should arrive at the functional relation

$$u/\sqrt{(\tau_\omega/\rho)} = \phi(y/\epsilon) \quad [3.3.22]$$

where ϵ is the equivalent size of the surface roughness.

Combining Eqn. [3.3.22] with the Prandtl velocity-defect law we arrive at the universal velocity profile for flow in a rough pipe

$$u/\sqrt{(\tau_\omega/\rho)} = A' + B' \log_{10}(y/\epsilon) \quad [3.3.23]$$

which is analogous to Eqn. [3.3.11] for a smooth pipe.

Good agreement is obtained with experimental measurements if we take $A' = 8{\cdot}5$ and $B' = 5{\cdot}75$ for the case of Newtonian fluids, i.e. Eqn. [3.3.23] becomes

$$u^+ = 8{\cdot}5 + 5{\cdot}75 \log_{10} (y/\epsilon) \quad [3.3.24]$$

Similarly it has been found that the turbulent friction factor for Newtonian fluids in rough pipes is given by

$$1/\sqrt{(c_f)} = 3{\cdot}46 + 4{\cdot}0 \log_{10} (a/\epsilon) \quad [3.3.25]$$

which is analogous to Eqn. [3.3.2] for smooth pipes.

Dodge and Metzner have suggested equivalent forms of Eqns. [3.3.24] and [3.3.25] for the case of non-Newtonian turbulent flow in rough pipes. The coefficients are now functions of n'. Although these coefficients have not been determined the important fact was emphasized that the constant 4·0 which appears in the relations for Newtonian fluids, hitherto regarded as a universal constant, is nothing more than a point value of a continuous function. The fact that the coefficient 4·0 appears in the equations for both rough and smooth pipes for Newtonian fluids is merely due to the special case that $n' = 1$. In the general case where n' is not equal to unity the coefficients in the equations for rough and smooth pipes must be expected to differ.

3.4 CRITERIA FOR THE ONSET OF TURBULENCE FOR NON-NEWTONIAN FLUID SYSTEMS

Various criteria for the onset of turbulence have been suggested by different workers, but as yet no reliable criterion has emerged. Winding, Baumann and Kranich [44] in their early work on pseudoplastic non-Newtonian systems suggested that the transition from laminar to turbulent flow in circular pipes occurred when the Reynolds number, defined as $\rho u_m D/\mu_0$, where μ_0 is the viscosity at zero shear (the slope of the tangent to the flow curve at the origin), reached a value of 2100. This is unconvincing since the viscosity at zero shear rate is unlikely to be a relevant factor at the high shear rates encountered in the turbulent state. Alves, Boucher and Pigford, [41] supported by Oldroyd, [46] suggested that turbulence occurs when the Reynolds number defined as $\rho u_m D/\mu_\infty$ reaches 2100; this appears more reasonable.

Hedström [34] applied dimensional analysis to the problem of turbulent flow of Bingham plastics in round pipes and plotted the friction factor as a function of $\rho u_m D/\mu_p$ and the Hedström group, $\tau_y D^2 \rho/\mu_p^2$, in the laminar region. The results in the turbulent region are plotted as a function of the Reynolds number only, implying that the contribution of the Hedström number is negligible in the turbulent zone. It was suggested that the onset of turbulence occurs when the laminar friction factor curve (a function of He) intersects the turbulent friction factor curve for Newtonian fluids (see Fig. 34). This implies that the critical Reynolds number increases as the yield stress, τ_y, increases, which is in accordance with experiment.

Weltmann's correlation (which was discussed in Section 3.1) is based on the integration of the power law for pseudoplastic and dilatant materials and in the laminar region it consists of a family of curves with the flow behaviour index, n, as parameter. This can be compared with Hedström's correlation for Bingham plastics which gives a family of curves in the laminar region with τ_y as parameter and Weltmann proposed a similar criterion for the onset of turbulence, namely, the Reynolds number given by the intersection of the line in the laminar region for the particular value of n in question and the turbulent friction factor line for Newtonian fluids which is appropriate for the pipe roughness. This is shown in Fig. 36.

Metzner and Reed [19] proposed a criterion for the onset of turbulence based on the friction factor and suggested that the transition from laminar to turbulent flow in round pipes occurred when the friction factor fell below 0·008. This criterion was used in conjunction with the generalized friction factor diagram given in Fig. 35.

More recently Dodge and Metzner[45] found that the critical generalized Reynolds number corresponding to the onset of turbulence appeared to increase with decreasing values of the flow behaviour index. At n' equal to unity (Newtonian case) the well-known value of the critical Reynolds number is 2100. This increased to about 3100 at a value of n' equal to 0·38. However no generalized criterion was established.

3.5 ENTRANCE LENGTHS AND EXPANSION AND CONTRACTION LOSSES

In pipe-line design the pressure losses in fittings such as valves, orifices, venturi sections, abrupt changes of diameter, etc., are of prime importance and often account for a considerable proportion of the total pressure loss in the complete system. For Newtonian flow the magnitude of losses of this kind are well known, and, although the theory is far from complete, the well-established design methods are usually quite satisfactory. Very few data are available for the flow of non-Newtonian fluids through fittings, but the indications are that the effects are different from the corresponding Newtonian behaviour. This is likely to be particularly true for viscoelastic materials.

(a) Entrance lengths in laminar flow

When a fluid flows into a pipe from an infinite reservoir, as in Fig. 39, the

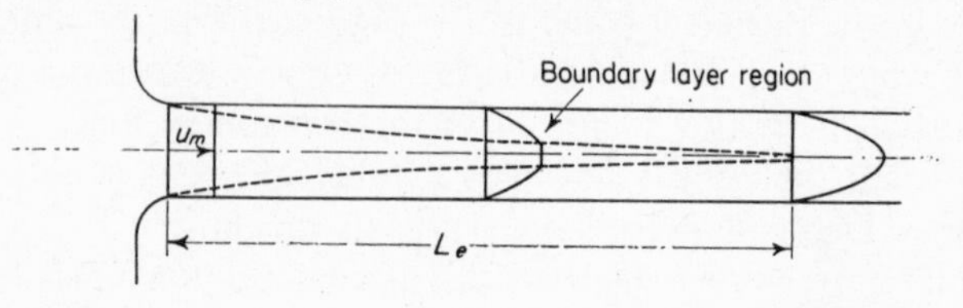

FIG. 39.

velocity at the entrance plane will be uniform across the pipe. By continuity this velocity will be equal to the mean velocity, u_m, for the fully developed profile some distance down the pipe. As the fluid moves down the pipe a boundary layer will form at the surface and this will gradually spread inwards towards the centre. The central core will be accelerated from the initial velocity u_m to the final velocity which will depend on the properties of the fluid. This will be equal to $2u_m$ for a Newtonian fluid and $u_m\,(3n + 1)/(n + 1)$ for a power law fluid. At this point the boundary layer will have converged at the centre line of the pipe and will now embrace the entire flow.

The distance from the entry at which this occurs is referred to as the *entrance length*. Within this region the pressure gradient will be greater than that for the final steady conditions owing to the changes which occur in the velocity profile, resulting in changes in the kinetic energy of the fluid stream and increased fluid friction.

For Newtonian fluids Schiller (47) has shown that the entrance length L_e is approximately given by

$$L_e/D = 0{\cdot}029 \text{ Re} \qquad [3.5.1]$$

where the Reynolds number is equal to $\rho u_m D/\mu$. This is found to be in good agreement with experiment.

Bogue (48) has recently extended this approach to include non-Newtonian fluids of the power law type. If we let the thickness of the boundary layer at a distance z be δ the following equation was derived:

$$\frac{z/a}{\text{Re}'} = \int_0^{\delta/a} \phi\left(\frac{\delta}{a}, n\right) \mathrm{d}\left(\frac{\delta}{a}\right) \qquad [3.5.2]$$

where ϕ is a function of δ/a and the flow behaviour index n, and the Reynolds number is in the generalized form for a power law fluid, i.e.

$$\text{Re}' = \frac{D^n u_m^{2-n} \rho}{\frac{k}{8}\left(\frac{6n+2}{n}\right)^n}$$

The entrance length L_e is given by the condition that the boundary layer reaches the centre line (i.e. $\delta = a$), giving from Eqn. [3.5.2]

$$\frac{L_e/a}{\text{Re}'} = \int_0^1 \phi\left(\frac{\delta}{a}, n\right) \mathrm{d}\left(\frac{\delta}{a}\right) \qquad [3.5.3]$$

Bogue has evaluated this integral as a function of the flow behaviour index n and plotted $L_e/a/\text{Re}'$ against n, indicating that the entrance length L_e increases with the Reynolds number (as in the case of Newtonian fluids) and decreases with n. The result reduces to the usual expression given in Eqn. [3.5.1] for the special case of a Newtonian fluid when $n = 1$ and $k = \mu$.

On qualitative grounds it may be argued that since the velocity profile for pseudoplastic fluids ($n < 1$) is flatter than that for Newtonian fluids, as shown in Fig. 37, the fully developed profile should be established more quickly and the entrance lengths for these fluids should be shorter, other things being equal. The reverse would be true for dilatant fluids which have a sharper velocity profile than Newtonian fluids. These conclusions are in accordance with the theoretical predictions of Bogue.

In turbulent flow the entrance lengths are considerably shorter than for laminar flow and nearly independent of the Reynolds number.

(b) Expansion losses in laminar flow

The approximate loss of head due to a sudden enlargement of a pipe can be calculated directly. We will assume that at section 1, Fig. 40, the fluid is in fully developed laminar flow and that following the expansion the flow will again become uniform at section 3. Immediately following the expansion at section 2 the streamlines will be assumed to be parallel and the pressure p_2 uniform across the pipe.

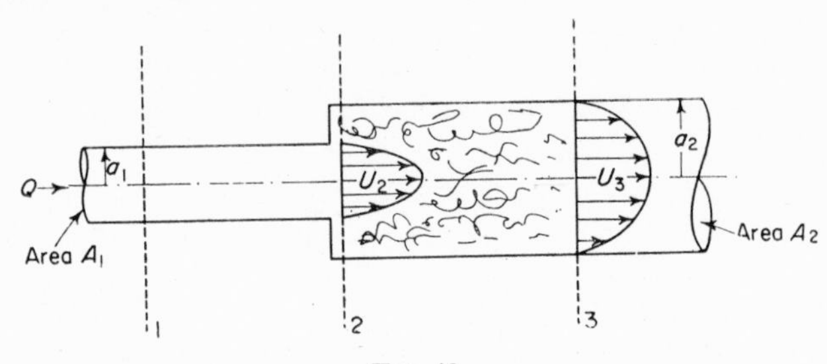

FIG. 40.

Neglecting friction between sections 1 and 2, the application of the Bernoulli equation along any streamline will give

$$p_1 = p_2 \qquad [3.5.4]$$

The fully developed velocity profile for a power law fluid is given by

$$u = U\,[1 - (r/a)^{(n+1)/n}] = u_m\,[(3n + 1)/(n + 1)]\,[1 - (r/a)^{(n+1)/n}] \qquad [3.5.5]$$

where U is the maximum velocity at the axis and u_m is the mean velocity. The momentum entering section 2 will be given by

$$M_2 = \int_0^{a_1} 2\pi r\, u_2^2\, \rho\, \mathrm{d}r$$

giving from Eqn. [3.5.5]

$$M_2 = \pi\rho a_1^2\, U_2^2\, [(n + 1)^2/(3\,n + 1)\,(2\,n + 1)] \qquad [3.5.6]$$

and similarly the momentum leaving section 3 will be

$$M_3 = \pi\rho a_2^2\, U_3^2\, [(n + 1)^2/(3\,n + 1)\,(2\,n + 1)] \qquad [3.5.7]$$

Applying the momentum principle between sections 2 and 3, neglecting skin friction at the wall of the pipe, the resultant force $(p_2 - p_3)\pi a_2^2$ will be equal to the net rate of outflow of momentum, i.e.

$$(p_2 - p_3)\pi a_2^2 = \frac{\pi\rho(n+1)^2}{(3n+1)(2n+1)}(U_3^2 a_2^2 - U_2^2 a_1^2) \qquad [3.5.8]$$

and since $U = Q(3n+1)/\pi a^2(n+1)$ and $p_1 = p_2$ we get

$$\frac{p_1 - p_3}{\rho} = \left(\frac{3n+1}{n+1}\right)\frac{Q^2}{A_1^2}\left[\left(\frac{A_1}{A_2}\right)^2 - \frac{A_1}{A_2}\right] \qquad [3.5.9]$$

where A_1 is the area at section 1 and A_2 the area at section 2.

The kinetic energy per unit time for fully developed laminar flow is given by

$$\text{K. E.} = \int_0^a \frac{1}{2}\, 2\pi r\ \rho\ u^3\, \mathrm{d}r$$

Substituting for u and integrating this becomes

$$\text{K. E.} = \frac{3\pi\rho U^3 a^2 (n+1)^3}{2(3n+1)(2n+1)(5n+3)}$$

Substituting for U in terms of Q gives

$$\text{K. E.} = \frac{3\pi\rho a^2 (3n+1)^2}{2(2n+1)(5n+3)}\left(\frac{Q}{\pi a^2}\right)^3 \qquad [3.5.10]$$

Dividing by ρQ we get the kinetic energy per unit mass

$$\text{K. E./unit mass} = \frac{3(3n+1)^2}{2(2n+1)(5n+3)}\left(\frac{Q}{A}\right)^2$$

The loss of mechanical energy per unit mass due to the contraction is therefore

$$\frac{p_1}{\rho} + \frac{3(3n+1)^2}{2(2n+1)(5n+3)}\left(\frac{Q}{A_1}\right)^2 - \frac{p^3}{\rho} - \frac{3(3n+1)^2}{2(2n+1)(5n+3)}\left(\frac{Q}{A_2}\right)^2$$

Dividing by g and substituting for $(p_1 - p_3)/\rho$ from Eqn. [3.5.9] we get the loss of head, i.e.

loss of head due to the sudden expansion =

$$\frac{1}{g}\left(\frac{Q}{A_1}\right)^2\left(\frac{3n+1}{2n+1}\right)\left[\frac{n+3}{2(5n+3)}\left(\frac{A_1}{A_2}\right)^2-\left(\frac{A_1}{A_2}\right)+\frac{3(3n+1)}{2(5n+3)}\right] \quad [3.5.11]$$

When $n = 0$ this reduces to

$$\text{loss of head} = \frac{u_1^2}{2g}\left(1-\frac{A_1}{A_2}\right)^2$$

which agrees with the approximate solution for *turbulent* flow of a Newtonian fluid when the velocity profile is assumed to be flat.

(c) Contraction losses

The above analysis cannot be applied in reverse to predict contraction losses in laminar flow because the size of the vena contracta is not known.

Most of the work on contraction losses is highly empirical and the findings of the various workers are usually at variance. Toms,[49] using non-Newtonian solutions of high polymers, concluded that the total contraction loss including the excess kinetic energy change and fluid friction was equal to ρu_m^2 for both laminar and turbulent flow following an infinite contraction. For laminar flow of Newtonian fluids the mean kinetic energy is given by

$$\frac{1}{\pi a^2 u_m}\int_0^a \frac{1}{2}\,2\pi r\rho u^3\,dr$$

which on substitution for u reduces to ρu_m^2, and this will be equal to the pressure drop associated with the acceleration of the fluid. Toms's conclusions suggest, then, that the excess fluid friction was negligible in laminar flow. His result is not easily justified for turbulent flow.

On the other hand Weltmann and Keller[50] found that there was no appreciable difference between non-Newtonian and Newtonian losses in either laminar and turbulent flow. This agrees with Toms's conclusions for laminar flow but not for turbulent flow.

Much higher losses are reported by McMillen[33] who studied a Bingham plastic gel in laminar flow and measured the pressure loss following sudden contractions from 2 to $\frac{3}{4}$ in. and from $\frac{3}{4}$ to $\frac{1}{8}$ in. at different flow rates. Contraction losses varied between $2\rho u_m^2$ and $5\rho u_m^2$ (where u_m is the mean velocity following the contraction) and the loss increased with the 1·5 power of the velocity.

The contraction loss (defined by extrapolating the pressure curve back to zero length as in Fig. 41)

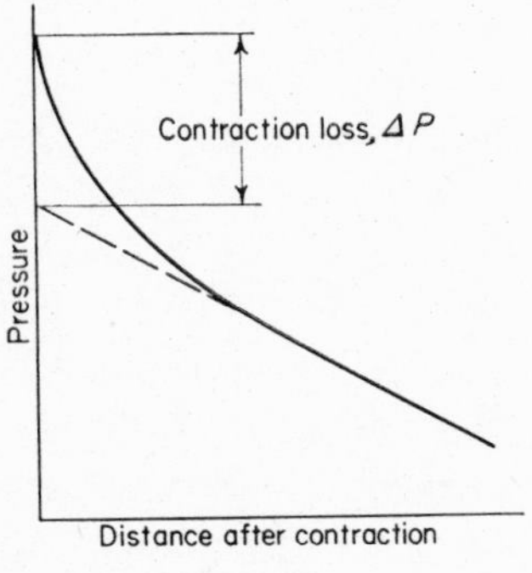

FIG. 41.

was correlated by

$$\underset{(\text{lb/in}^2)}{\Delta P} = 0{\cdot}12 \underset{(\text{ft/sec})}{u_m^{1\cdot5}}$$

Recent preliminary investigations by Dodge,[45] however, do not support these conclusions, emphasizing once again that at the present time there is a severe lack of reliable information in this field.

3.6 LAMINAR AXIAL FLOW OF A NON-NEWTONIAN FLUID IN AN ANNULUS

The flow of non-Newtonian fluids through an annulus is a problem of considerable industrial importance. One notable example is in oil well drilling where a heavy drilling mud is circulated through the annular space round the drill pipe in order to carry the drillings to the surface. These drilling muds are usually highly non-Newtonian, either Bingham plastic or pseudoplastic in character.

In the manufacture of extruded plastic tubes the polymer melt, usually non-Newtonian, is forced by the extruder screw through an annular die, and here again it is important to be able to predict the relation between pressure and throughput for the system. Another example of the same basic problem occurs in parallel-tube heat exchangers which have to handle non-Newtonian fluids. Here the problem is complicated by the fact that the flow is not isothermal.

(a) Basic equations

Consider the balance of forces on an element of axial length δL between radii r and $r + \delta r$ as shown in Fig. 42.

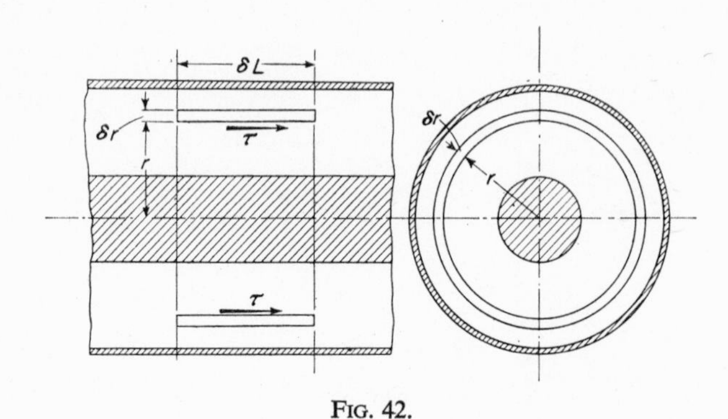

FIG. 42.

This gives

$$2\pi r \; \delta r \; \delta P = 2\pi \delta L \frac{\mathrm{d}}{\mathrm{d}r} (\tau r) \; \delta r \qquad [3.6.1]$$

where τ is the shear stress at radius r and δP is the pressure drop over length δL, i.e. $\mathrm{d}P/\mathrm{d}L$ is the pressure gradient in the annulus.

Integrating Eqn. [3.6.1] we get

$$\tau r = \tfrac{1}{2} r^2 \mathrm{d}P/\mathrm{d}L + \text{constant} \qquad [3.6.2]$$

If $\tau = 0$ when $r = \lambda R$, i.e. at the radius of maximum velocity, the constant in Eqn. [3.6.2] becomes

$$-\tfrac{1}{2}(\lambda R)^2 \mathrm{d}P/\mathrm{d}L$$

hence

$$\tau = \tfrac{1}{2}(\mathrm{d}P/\mathrm{d}L)\,[r - (\lambda R)^2/r] \qquad [3.6.3]$$

This is the basic equation which must be solved, subject to the boundary conditions of zero velocity at the walls and the rheological equation of the fluid, $\tau = f(\dot{\gamma})$.

(b) Solution for a Bingham plastic fluid

Fredrickson and Bird [51] have solved this problem and an outline of their solution will be given here. For details the reader is referred to the original paper.

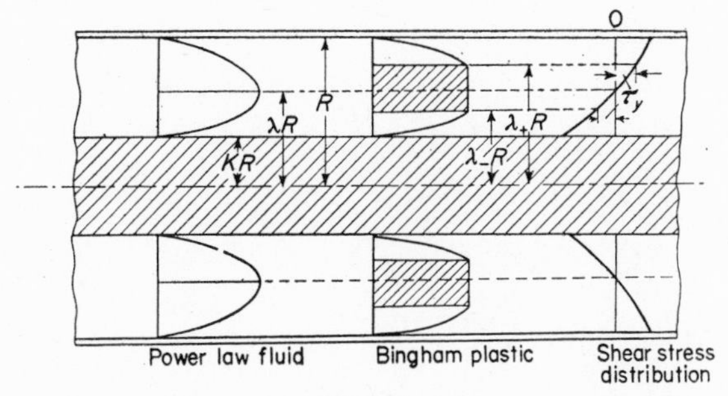

FIG. 43.

The nomenclature is given in Fig. 43. λ_+R and λ_-R define the outer and inner radii of the solid core, i.e. those radii at which the local shear stress is equal to the yield value of the material; κR is the radius of the inner cylinder.

The solution is considerably simplified by introducing the following dimensionless variables:

$$\left.\begin{aligned} T &= \frac{2\tau}{R\mathrm{d}P/\mathrm{d}L} && \text{dimensionless shear stress} \\ T_0 &= \frac{2\tau_y}{R\mathrm{d}P/\mathrm{d}L} && \text{dimensionless yield stress} \\ \phi &= \frac{2\mu_p}{R^2\mathrm{d}P/\mathrm{d}L}u && \text{dimensionless velocity} \\ \rho &= r/R && \text{dimensionless radial distance} \end{aligned}\right\} \quad [3.6.4]$$

It can then be shown that the dimensionless velocity for the region between the core and the inner cylinder, i.e. $\kappa \leqslant \rho \leqslant \lambda_-$, is given by

$$\phi_- = T_0(\kappa - \rho) - \tfrac{1}{2}(\rho^2 - \kappa^2) + \lambda^2 \ln(\rho/\kappa) \quad [3.6.5]$$

Similarly for the region between the solid core and the outer wall, i.e. for $\lambda_+ \leqslant \rho \leqslant 1$, we have

$$\phi_+ = T_0(\rho - 1) + \tfrac{1}{2}(1 - \rho^2) + \lambda^2 \ln \rho \quad [3.6.6]$$

In the solid core, i.e. in the region $\lambda_- \leqslant \rho \leqslant \lambda_+$, we have

$$\phi_0 = \phi_-(\lambda_-) = \phi_+(\lambda_+) \quad [3.6.7]$$

Eqn. [3.6.7] determines λ_+ and ϕ_0 as functions of κ and T_0. Fredrickson and Bird have calculated these functions and presented the results graphically as plots of λ_+ and ϕ_0 versus κ for various values of T_0. The result is shown in Fig. 44.

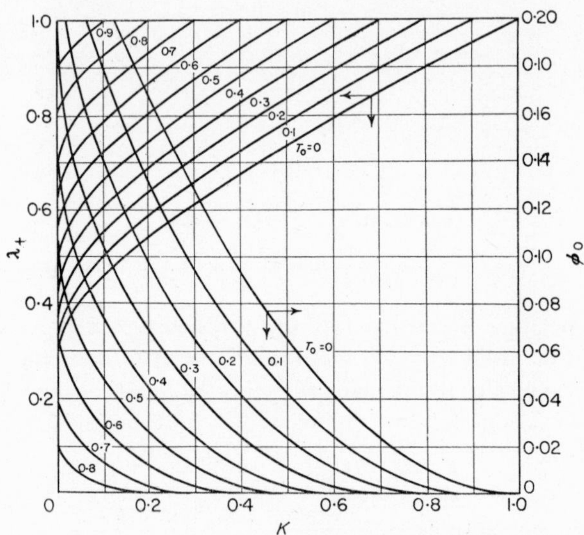

(By courtesy of *Industrial and Engineering Chemistry*.)

FIG. 44.

λ_+ defines the outer radius of the plug. The inner radius then follows since

$$\lambda_- = (\lambda_+ - T_0) \quad [3.6.8]$$

also

$$\lambda^2 = \lambda_+ \lambda_- \quad [3.6.9]$$

The total volumetric flow rate is given by

$$Q = 2\pi R^2 \int_{\kappa}^{1} u\rho \, \mathrm{d}\, \rho$$

and the final result is

$$Q = \frac{\pi R^4 \mathrm{d}P/\mathrm{d}L}{8\mu_p}\Big[(1 - \kappa^4) - 2\lambda_+(\lambda_+ - T_0)(1 - \kappa^2) - \frac{4}{3}(1 + \kappa^3)T_0 + \frac{1}{3}(2\lambda_+ - T_0)^3\, T_0\Big] \quad [3.6.10]$$

The term in the square bracket of Eqn. [3.6.10] is a function of κ, λ_+ and T_0 and hence a function of κ and T_0 since λ_+ is also a function of κ and T_0. Hence

$$Q \bigg/ \frac{\pi R^4 \frac{dP}{dL}}{8\mu_p} = \Omega_B(\kappa, T_0) \qquad [3.6.11]$$

Fredrickson and Bird have presented the function Ω_B graphically and the result is shown in Fig. 45.

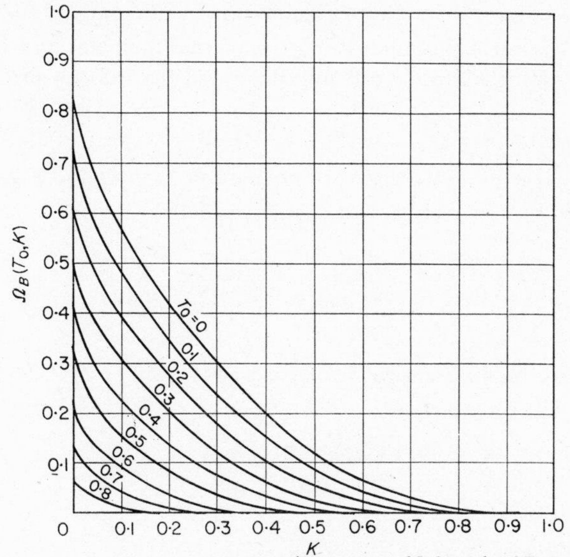

(By courtesy of *Industrial and Engineering Chemistry*.)

FIG. 45.

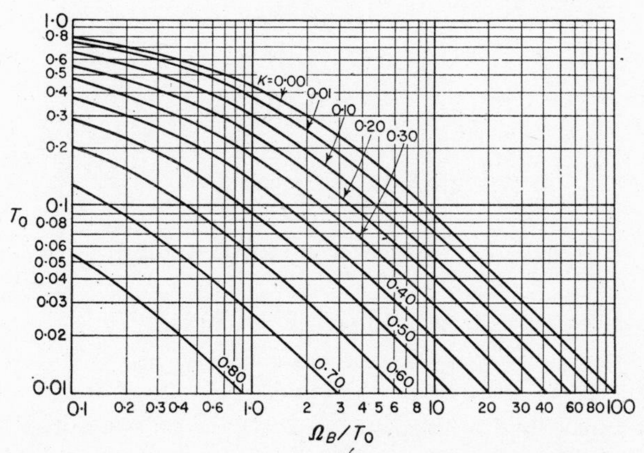

(By courtesy of *Industrial and Engineering Chemistry*.)

FIG. 46.

Fig. 45 enables the flow rate, Q, to be found if the pressure gradient, the dimensions of the annulus, and the properties of the fluid, τ_y and μ_p, are known.

For calculating the pressure gradient for a given flow rate a plot of the function Ω_B/T_0 is more convenient. This is given in Fig. 46.

*Example**

A Bingham plastic drilling mud with a yield stress of 0·15 lb/ft² and a plastic viscosity of $4{\cdot}178 \times 10^{-4}$ lb sec/ft² flows at the rate of 4·5 ft³/sec in an annulus of 8 in. internal diameter, 12 in. external diameter and 1000 ft long. It is required to calculate the pressure drop and the velocity profile.

(*i*) *Pressure drop*

To calculate the pressure drop at a given flow rate we use Fig. 46.

From Eqns. [3.6.11] and [3.6.4] for T_0 we have

$$\frac{\Omega_B}{T_0} = \frac{8\mu_p Q}{\pi R^4 \mathrm{d}P/\mathrm{d}L} \frac{R\mathrm{d}P/\mathrm{d}L}{2\tau_y} = \frac{4\mu_p Q}{\pi R^3 \tau_y}$$

Hence substituting numerical values we get

$$\frac{\Omega_B}{T_0} = \frac{4(4{\cdot}178 \times 10^{-4})(4{\cdot}5)}{\pi(0{\cdot}125)(0{\cdot}15)} = 0{\cdot}128$$

The ratio of the radii $= 8/12 = 0{\cdot}667$.

From Fig. 46 it is seen that $T_0 = 0{\cdot}140$ when $\Omega_B/T_0 = 0{\cdot}128$ and $\kappa = 0{\cdot}667$.

But
$$T_0 = \frac{2\tau_y}{R\,\mathrm{d}P/\mathrm{d}L}$$

i.e.
$$\mathrm{d}P/\mathrm{d}L = 2\tau_y/RT_0$$

$$=(2 \times 0{\cdot}15)/(0{\cdot}5 \times 0{\cdot}140) = 4{\cdot}3 \text{ lb/ft}^3$$

Hence the total pressure drop is 4300 lb/ft² or 30 lb/in².

Note

For a Newtonian fluid of the same viscosity the pressure drop would have been 12·9 lb/in².

*Fluid properties taken from Laird, W. M., Ind. Eng. Chem. 1957 **49** 138.

(*ii*) *Velocity profile*

From Fig. 44 at $\kappa = 0{\cdot}667$ and $T_0 = 0{\cdot}140$ we have

$$\phi_0 = 0{\cdot}01$$

and

$$\lambda_+ = 0{\cdot}90$$

From the definition of ϕ in Eqn. [3.6.4] the velocity of the plug, u_p, is given by

$$u_p = \frac{\phi_0 R^2 \, \mathrm{d}P/\mathrm{d}L}{2\mu_p}$$

$$= (0{\cdot}01 \times 0{\cdot}25 \times 4{\cdot}3)/(2 \times 4{\cdot}178 \times 10^{-4})$$

i.e.

$$u_p = 13 \text{ ft/sec}$$

Since $\lambda_+ = 0{\cdot}90$ and $T_0 = 0{\cdot}140$ we have from Eqn. [3.6.8]

$$\lambda_- = (0{\cdot}9 - 0{\cdot}140) = 0{\cdot}760$$

Hence from Eqn. [3.6.9]

$$\lambda = \sqrt{(\lambda_+ \lambda_1)} = 0{\cdot}830$$

Hence outer radius of plug $= \lambda_+ R = 0{\cdot}9 \times 0{\cdot}5 = 0{\cdot}45$ ft and inner radius of plug $= \lambda_- R = 0{\cdot}760 \times 0{\cdot}5 = 0{\cdot}38$ ft.

Knowing the values of T_0, κ and λ the velocity profiles outside the solid plug may be calculated from Eqns. [3.6.5] and [3.6.6]. The result is shown in Fig. 47.

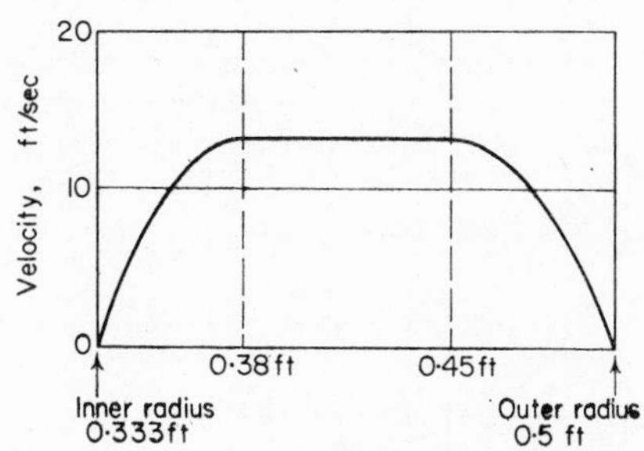

(By courtesey of *Industrial and Engineering Chemistry*.)

FIG. 47.

This result agrees with that of Laird who used a different method of analysis.

(c) Solution for a power law fluid

For a power law fluid the rheological equation is

$$\tau = - k \, (\mathrm{d}u/\mathrm{d}r)^n \qquad [3.6.12]$$

By combining Eqns. [3.6.12] and [3.6.3] and integrating it may be shown that[51] the velocity distribution is given by

$$u = R \left(\frac{\mathrm{d}P}{\mathrm{d}L} \cdot \frac{R}{2k}\right)^S \int_{\kappa}^{\rho} \left(\frac{\lambda^2}{\rho} - \rho\right)^S \mathrm{d}\rho \;; \kappa \leqslant \rho \leqslant \lambda \qquad [3.6.13]$$

and

$$u = R \left(\frac{\mathrm{d}P}{\mathrm{d}L} \cdot \frac{R}{2k}\right)^S \int_{\rho}^{1} \left(\rho - \frac{\lambda^2}{\rho}\right)^S \mathrm{d}\rho \;; \lambda \leqslant \rho \leqslant 1 \qquad [3.6.14]$$

if $u = 0$ at $\rho = \kappa$ and $\rho = 1$, where $S = 1/n$. When $\rho = \lambda$ Eqns. [3.6.13] and [3.6.14] must give the same value of u, i.e.

$$\int_{\kappa}^{\lambda} (\lambda^2/\rho - \rho)^S \, \mathrm{d}\,\rho = \int_{\lambda}^{1} (\rho - \lambda^2/\rho)^S \, \mathrm{d}\,\rho \qquad [3.6.15]$$

This equation determines λ as a function of κ and S, and the result is shown graphically in Fig. 48.

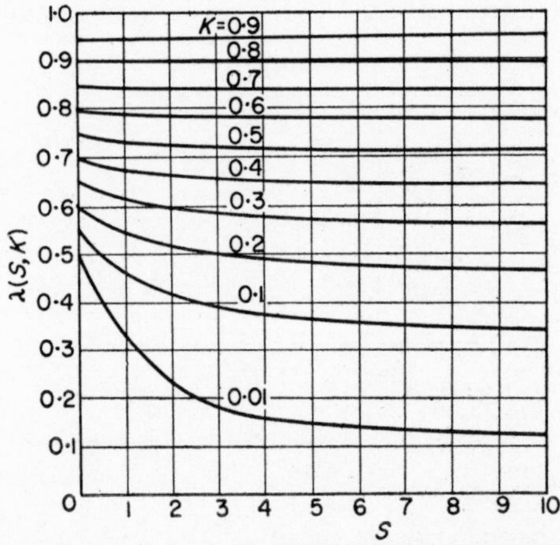

(By courtesy of *Industrial and Engineering Chemistry.*)

FIG. 48.

The volume rate of flow is given by

$$Q = 2\pi R^2 \int_\kappa^1 u \ \rho d\rho \qquad [3.6.16]$$

Substituting for u from Eqns. [3.6.13] and [3.6.14] into [3.6.16] we finally obtain

$$Q = \pi R^3 \left(\frac{dP}{dL} \cdot \frac{R}{2k}\right)^S \int_\kappa^1 \left|\lambda^2 - \rho^2\right|^{S+1} \rho^{-S} \, d\rho \qquad [3.6.17]$$

This may be written as

$$Q = \pi R^3 \left(\frac{dP}{dL} \frac{R}{2k}\right)^S \Omega_p (S, \kappa) \qquad [3.6.18]$$

since the integral in Eqn. [3.6.17] is a function of S and κ.

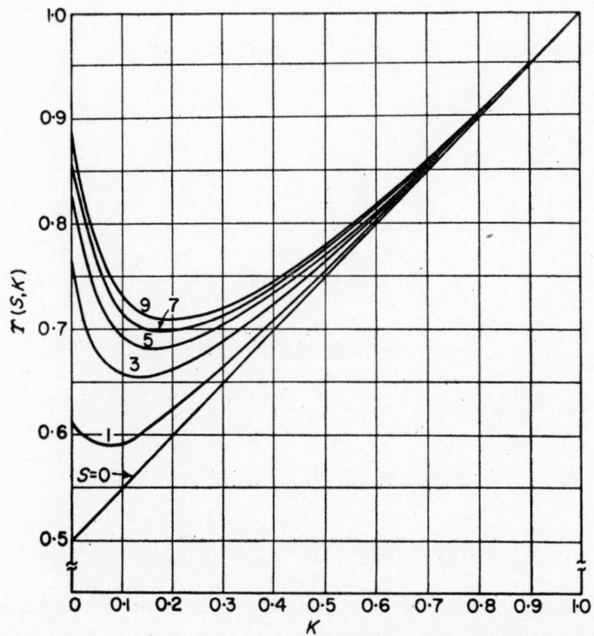

(By courtesy of *Industrial and Engineering Chemistry*).

FIG. 49.

Fredrickson and Bird have presented the function $\Omega_p(S, \kappa)$ graphically by plotting the function $\gamma(S, \kappa)$, defined as

$$\gamma(S, \kappa) = (S + 2)\, \Omega_p(S, \kappa)/(1 - \kappa)^{S+2}$$

as a function of S and κ. This is shown in Fig. 49.

This enables the throughput for any pressure drop to be calculated if the dimensions of the annulus, R and κ, and the properties of the fluid, k and S, are known.

Example

A fluid for which $S = 1{\cdot}398$ and $k = 0{\cdot}00635$ lb. sec$^{0{\cdot}716}$/ft^2 flows through an annulus of outside radius 1·033 in. and inside radius 0·420 in. at a rate of 0·098 ft^3/sec. It is required to calculate the pressure drop per unit length.

$$\kappa = 0{\cdot}406$$

$$R = 0{\cdot}0863 \text{ ft}$$

From Fig. 49 at $\kappa = 0{\cdot}406$ and $S = 1{\cdot}398$ we have

$$\gamma(S, \kappa) = 0{\cdot}72$$

But

$$\Omega_p = \frac{(1 - \kappa)^{S+2}\gamma(S, \kappa)}{(S + 2)} = \frac{(1 - 0{\cdot}406)^{3{\cdot}398}\,(0{\cdot}72)}{3{\cdot}398}$$

or

$$\Omega_p = 0{\cdot}0359$$

Also

$$Q = \pi R^3 (\mathrm{d}P/\mathrm{d}L)^S\,(R/2k)^S\,\Omega_p$$

and after substituting for Q, R, k and Ω_p we get

$$\mathrm{d}P/\mathrm{d}L = 0{\cdot}177 \text{ lb/in}^2 \text{ per foot length.}$$

3.7 EXTRUSION OF POLYMER MELTS

During the past few years many papers have appeared which have attempted to derive theoretically the relations which govern the behaviour of screw extruders. Most of the early work was concerned with the isothermal or adiabatic extrusion of polymers which were assumed to behave as viscous Newtonian fluids under the conditions existing in the extruder. These studies were presented in an extrusion symposium in 1953.[52]

It is unlikely that the assumption of Newtonian behaviour is a good one in practice for many of the polymer melts which have to be processed, and recently attention has been focussed on the derivation of equations based on an assumed rheological equation which approximates the true behaviour more closely.

The basic equations for Newtonian fluids will be derived and the extensions to non-Newtonian systems will then be considered.

(a) Newtonian fluids

Consider the diagram of a screw extruder given in Fig. 50 which shows the notation to be used.

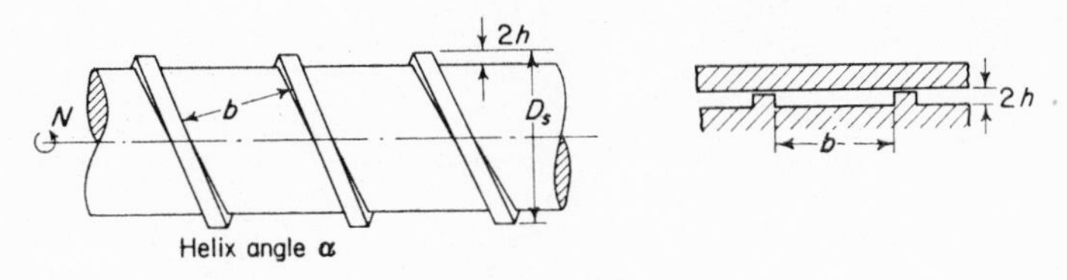

FIG. 50.

Although it is the screw which rotates in practice it is easier to imagine the flow situation by considering the barrel rotating round the stationary screw. If the curvature of the channel is neglected the system is equivalent to flow between parallel plates, one of which is stationary and the other moving with a velocity equal to the peripheral speed of the barrel resolved along the channel. There will also be an adverse pressure gradient along the channel, dP/dz. This simplified picture is shown in Fig. 51.

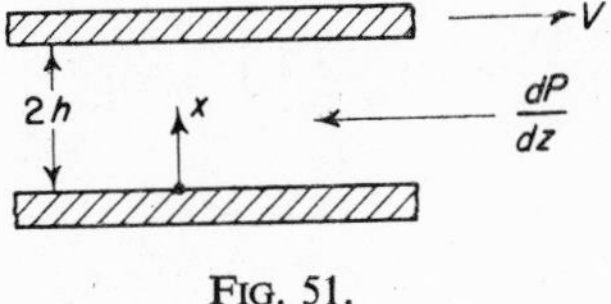

FIG. 51.

Neglecting end effects (since $b \gg 2h$) and assuming no slip at the walls the velocity distribution will be given by

$$v(x) = \frac{Vx}{2h} + \frac{x^2 - 2hx}{2\mu} \frac{dP}{dz} \qquad [3.7.1]$$

where μ is the Newtonian viscosity and $\mathrm{d}P/\mathrm{d}z$ is the pressure gradient along the channel.

The flow rate is given by

$$Q = \int_0^{2h} b\, v\, \mathrm{d}x$$

which reduces to

$$\frac{Q}{bhV} = 1 - \frac{2h^2}{3\mu V}\frac{\mathrm{d}P}{\mathrm{d}z} \qquad [3.7.2]$$

This can be shown graphically as in Fig. 52.

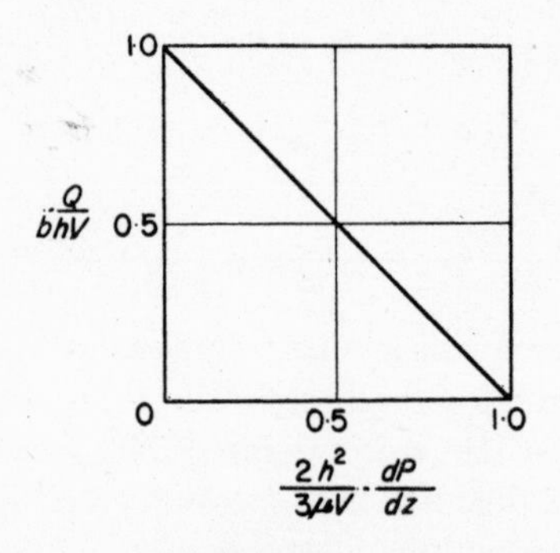

FIG. 52.

The velocity profile depends on the magnitude of $\mathrm{d}P/\mathrm{d}z$. There are three cases

(*i*) $Q/bhV = 1$; $\mathrm{d}P/\mathrm{d}z = 0$

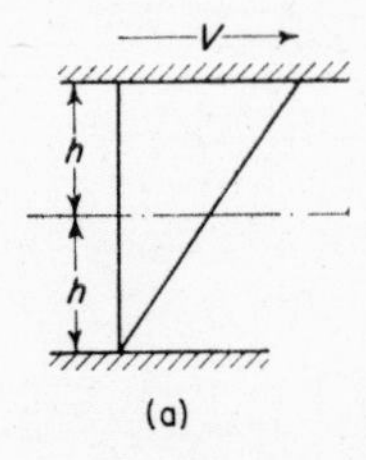

FIG. 53(a).

(*ii*) $Q/bhV < 1$; dP/dz is positive

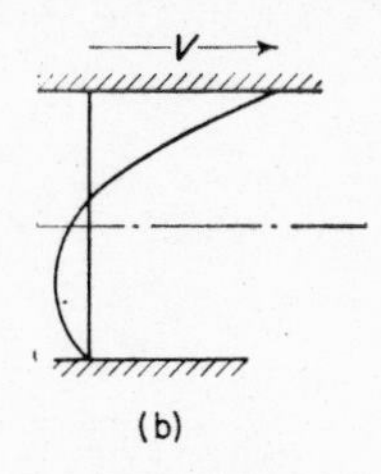

FIG. 53(b).

(*iii*) $Q/bhV > 1$; dP/dz is negative

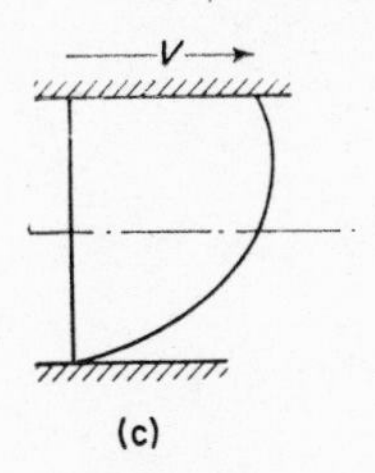

FIG. 53(c)

Case (*ii*) is the normal state of affairs in extrusion through a die. Case (*i*) would occur with no die.

(b) Fluids obeying the Rabinowitsch equation

Polyethylene and polystyrene melts can be described approximately by a rheological equation proposed by Rabinowitsch [53] which takes the form

$$\dot{\gamma} = \frac{\tau}{\mu_0}(1 + c\,\tau^2) \qquad [3.7.3]$$

where c and μ_0 are constants, typical of the material.

Yoshida *et al.* [54] have derived the equations which govern the extrusion of materials which can be described by this rheological equation. The velocity profile is given by

$$v = (V/2) + (\xi/2\mu_0)[y^2 - h^2(1 + \lambda^2)] + (c\xi^3/4\mu_0)[y^4 - h^4(1 + 6\lambda^2 + \lambda^4)] \qquad [3.7.4]$$

where for convenience the pressure gradient dP/dz is written as ξ, $\lambda = S/h$, and y and S are defined in Fig. 54.

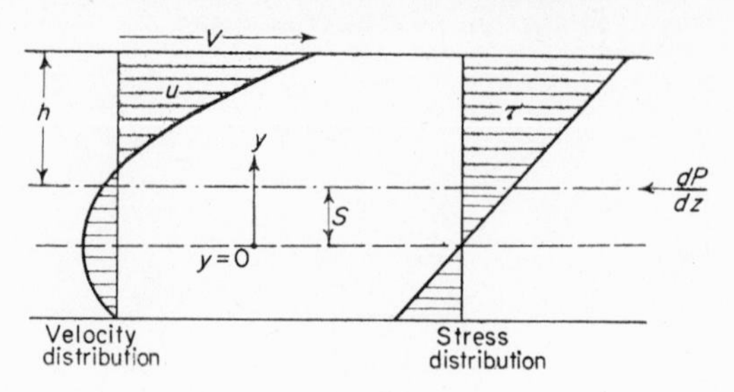

FIG. 54.

The defining equation for λ is

$$V = (2h^2\lambda/\mu_0)[\xi + ch^2\xi^3(1 + \lambda^2)] = \pi D_S N \cos \alpha \qquad [3.7.5]$$

and the volumetric flow rate is given by

$$Q = bh[V - (2h^2\xi/3\mu_0) - (2ch^4\xi^3/5\mu_0)(1 + 5\lambda^2)] \qquad [3.7.6]$$

The authors considerably simplified the use of these equations by introducing the following dimensionless groups,

$$u = Q/bhV \; ; \; \omega = \frac{V\mu_0\sqrt{c}}{h} \; ; \; z = \xi h\sqrt{c} \qquad [3.7.7]$$

and the results were presented graphically by plotting z against λ with contours of constant u and ω. At any given cross-section of the screw channel the values of b, h, c and μ_0 are known. Hence the values of u and ω can be calculated for constant mass flow rate Q_m (equal to ρQ) and constant V (or N). The values of z and λ are then read from the graph and from z the pressure gradient, ξ, is calculated. Thus for any constant speed the pressure gradient in the channel can be plotted as a function of channel length for various values of the mass throughput, Q_m. The area under this curve at any value of the channel length is $\int \frac{dP}{dz}\,dz$, i.e. the pressure in the channel at this point. This gives the pressure distribution in the extruder. Experiments were carried out on polyethylene in the temperature range 130—200°C and the results were found to agree very well with the theoretical predictions.

It should be noted here that apart from taking into account the non-Newtonian behaviour of the polymer this method can allow for any desired temperature distribution along the extruder.

The condition for a Newtonian fluid is that $c = 0$ in Eqn. [3.7.3]. Substituting $c = 0$ into Eqn. [3.7.6] gives

$$Q/bhV = 1 - 2h^2\xi/3\mu_0 V$$

and since $\xi = \mathrm{d}P/\mathrm{d}z$ this agrees with Eqn. [3.7.2].

(c) Power law fluid

Mori and Matsumoto [55] have recently presented the equations for the extrusion of materials which can be described by a power law of the form

$$\dot{\gamma} = \tau^P/\mu \qquad [3.7.8]$$

The equations, analogous to Eqns. [3.7.5] and [3.7.6] of the previous section, are

$$\frac{Q}{bhV} = (\lambda + 1) - \frac{(P + 1)}{(P + 2)}\left[\frac{(\lambda + 1)^{P+2} - (\lambda - 1)^{P+2}}{(\lambda + 1)^{P+1} - (\lambda - 1)^{P+1}}\right] \qquad [3.7.9]$$

and

$$\frac{V\mu}{h} = \frac{1}{\mathrm{P} + 1}\left(\frac{\xi}{h}\right)^P \left[(\lambda + 1)^{P+1} - (\lambda - 1)^{P+1}\right] \qquad [3.7.10]$$

where $\lambda = S/h$ as before and μ is defined by Eqn. [3.7.8]. The pressure gradient at any point in the channel can be calculated from these equations and the known temperature distribution. The pressure distribution then follows by integration. This was done and compared with experiment and the agreement was very close.

(d) Bingham plastics

Mori and Otatake [56] have solved the equations for the case of fluids which may be assumed to be Bingham plastic in behaviour. The results are presented graphically in terms of four dimensionless groups

$$Q/bhV;\ S/h;\ \mu_P V/\tau_y h;\ \tau_y/h\xi \qquad [3.7.11]$$

where μ_P is the Bingham plastic viscosity, τ_y is the yield stress and the other terms have the same meaning as before.

This approach is not as useful as the previous two since it is unlikely that many plastics will follow the Bingham equation closely.

(e) Arbitrary fluid properties

Recently, Nickolls and Colwell [57] presented a paper which described a method for estimating extruder performance when the screw and barrel surfaces are maintained at arbitrary temperatures and the fluid deviates from Newtonian behaviour in an arbitrary manner.

The method has been programmed for machine computation and typical results are presented. A graphical solution is also discussed.

3.8 ROLLING OF PLASTICS

In many of the plastics and allied industries sheets are formed by rolling a slab of material between cylindrical rolls and the material is usually non-Newtonian in character. Gaskell [58] has presented the following mathematical description of this process.

The process is illustrated in Fig. 55. The material is compressed and forced through the nip of the rolls, the minimum clearance being $2t_0$. The

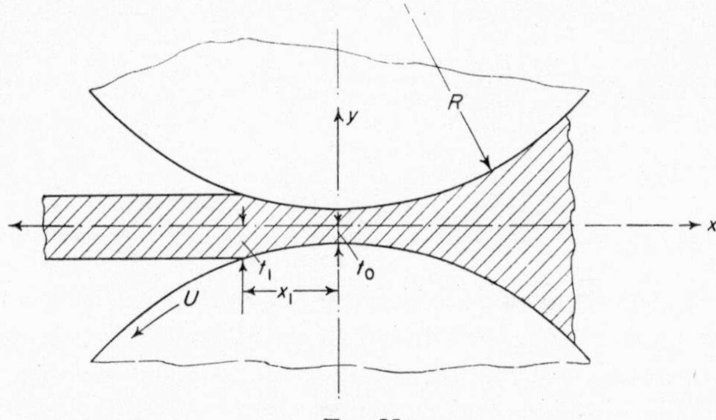

FIG. 55.

material leaves the rolls as a sheet some distance after the minimum clearance has been passed and not at the point of minimum clearance, as is often assumed for inelastic materials. Gaskell has explained the excess thickness over the thickness of the nip on the basis of the flow characterstics of the material and not as a consequence of elasticity.

It will be assumed that the rate of shear of the material is a function of the shear stress alone, i.e.

$$du/dy = f(\tau) \qquad [3.8.1]$$

By considering the equilibrium of a small element between y and $y + \delta y$ and x and $x + \delta x$ we find that the pressure gradient is given by

$$dp/dx = d\tau/dy \qquad [3.8.2]$$

In order to simplify the equations the following dimensionless variables are introduced:

$$\left.\begin{aligned} \xi &= x/\sqrt{(2R\,t_0)} \\ \eta &= y/\sqrt{(2R\,t_0)} \\ \delta &= \sqrt{(2t_0/R)} \ll 1 \end{aligned}\right\} \qquad [3.8.3]$$

Substituting these into Eqns. [3.8.1] and [3.8.2] we get

$$\mathrm{d}u/\mathrm{d}\eta = UF(\tau) \qquad [3.8.4]$$

and

$$\mathrm{d}\tau/\mathrm{d}\eta = \mathrm{g}(\xi) \qquad [3.8.5]$$

where

$$F(\tau) = (2t_0/U\delta)\,f(\tau) \qquad [3.8.6]$$

and

$$g = \mathrm{d}p/\mathrm{d}\xi \qquad [3.8.7]$$

By integrating Eqn. [3.8.5] noting that τ is zero at the centre line we get

$$\tau = g\eta$$

Hence from Eqn. [3.8.4]

$$\mathrm{d}u/\mathrm{d}\eta = UF(g\,\eta) \qquad [3.8.8]$$

Integrating we get

$$u/U = 1 - (1/g)\int_{\tau}^{z} F(\beta)\mathrm{d}\,\beta \qquad [3.8.9]$$

where β is merely a variable of integration and z is the value of $g\eta$ at the surface of the roll. This can be found since by geometry we have that at the surface

$$y = \pm\,[t_0 + R - \sqrt{(R^2 - x^2)}]$$

which on expansion gives

$$y = \pm\,(t_0 + x^2/2R) \text{ since } x \ll R$$

Substituting for y and x from Eqns. [3.8.3] we get that

$$z = g\eta = g\delta\,(1 + \xi^2)/2 \qquad [3.8.10]$$

Integration of Eqn. [3.8.9] gives the rate of flow

$$Q = \frac{2UR\delta}{g}\int_{0}^{z}\left[1 - \frac{1}{g}\int_{\beta'}^{z} F(\beta)\mathrm{d}\,\beta\right]\mathrm{d}\,\beta' \qquad [3.8.11]$$

If $2t_1$ is the thickness at the exit as in Fig. 55 we have that

$$Q = 2Ut_1 = 2Ut_0(1 + \xi_1^2) \qquad [3.8.12]$$

and substituting for Q into Eqn. [3.8.11] gives

$$g^2\delta(\xi^2 - \xi_1^2)/2 = z\int_0^z F(\beta)\mathrm{d}\beta - \int_0^z\int_0^{\beta'} F(\beta)\mathrm{d}\beta\,\mathrm{d}\beta' \qquad [3.8.13]$$

The double integral can be written as

$$-\int_0^z F(\beta)(z - \beta)\mathrm{d}\beta$$

and Eqn. [3.8.13] then reduces to

$$2(\xi^2 - \xi_1^2)/\delta(1 + \xi^2)^2 = z^{-2}\int_0^z \beta F(\beta)\mathrm{d}\beta \qquad [3.8.14]$$

Eqn. [3.8.14] gives the relation between z and ξ and hence between g and ξ. Knowing ξ_1 from Eqn. [3.8.12] using the known values of roll speed and the flow rate of the material, the pressure gradient $g(\xi)$ can be computed as a function of ξ and hence as a function of the distance through the rolls. For a Newtonian fluid Gaskell has derived the following result for the pressure distribution

$$p = \frac{3\mu U}{4t_0\delta}\left\{\left[\frac{\xi^2 - 1 - 5\,\xi_1^2 - 3\,\xi^2\,\xi_1^2}{(1 + \xi^2)^2}\right]\xi + (1 - 3\xi_1^2)\tan^{-1}\xi\right\} + C$$

where the constant C is given by

$$C = (1 - 3\,\xi_1^2)\tan^{-1}\xi_1 - (1 + 3\xi_1^2)\xi_1/(1 + \xi_1^2) \qquad [3.8.15]$$

For a Bingham plastic the function $f(\tau)$ is given by

$$f(\tau) = 0\ ;\ 0 < \tau < \tau_y$$

$$\text{and } f(\tau) = (\tau - \tau_y)/\mu_p\ ;\ \tau_y < \tau$$

For this material it can be shown that Eqn. [3.8.14] reduces to

$$\frac{t_0}{6\mu_p U z^2}(2z^3 - 3\tau_y z^2 + 3\tau_y^{z} \pm 2\tau_y) = (\xi^2 - \xi_1^2)/(1 + \xi^2)^2 \qquad [3.8.16]$$

where the minus sign holds for $z \geqslant \tau_y$ and the plus sign for $z \leqslant -\tau_y$.

Since $\tau = 0$ at the centre line there will be a region near the centre line where the stress is less than the yield stress and the plastic will move as a solid plug. The thickness of the plug will depend on τ_y. Let it be given by η_0 for which $\tau = \tau_y$. Putting $\tau = g\eta_0 = \tau_y$ in Eqn. [3.8.16] we obtain

$$\frac{\tau_y t_0}{\mu_p} [3\rho^2 - 3 \pm 2(\rho^3 - 1/\rho)]/6U = (\xi^2 - \xi_1^2)/(1 + \xi^2)^2 \qquad [3.8.17]$$

where

$$\rho = 2\eta_0/\delta(1 + \xi^2)$$

Eqn. [3.8.17] expresses the thickness of the solid plug (in terms of η_0) as a function of the distance through the rolls (in terms of ξ). This solid region would be as shown in Fig. 56.

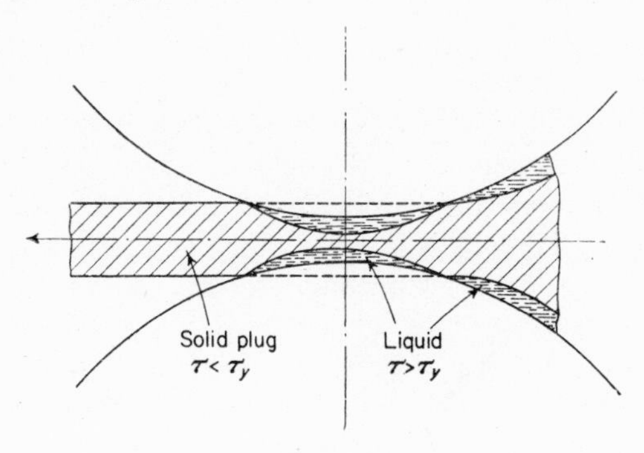

FIG. 56.

CHAPTER 4

HEAT-TRANSFER CHARACTERISTICS OF NON-NEWTONIAN FLUIDS

At the present time there is very little published information on the heat-transfer characteristics of non-Newtonian systems and the only problem which has received attention is the flow inside circular tubes.

Heat transfer to non-Newtonian fluids in laminar flow in tubes has been treated theoretically by a number of workers but there is a distinct lack of published experimental data which would allow these theoretical predictions to be fully tested.

The problem of the heat transfer to highly non-Newtonian fluids under turbulent flow conditions has received even less attention and so far only order of magnitude predictions are possible. However, this region is not as important as the laminar flow region for most non-Newtonian fluid systems. The consistency of these fluids is usually high and turbulent flow conditions, while desirable from heat transfer considerations, are often difficult to achieve in practice.

The present state of the subject will be discussed in the following sections.

4.1 HEAT TRANSFER IN LAMINAR FLOW IN A PIPE

Consider the case of steady-state laminar flow in a circular pipe with heat transfer. At the entry to the pipe let the fluid temperature be constant at T_1. Let the pipe temperature be constant at T_0 over the heated (or cooled) section and let the temperature at any other point be T. It will be assumed that the fluid properties are independent of temperature.

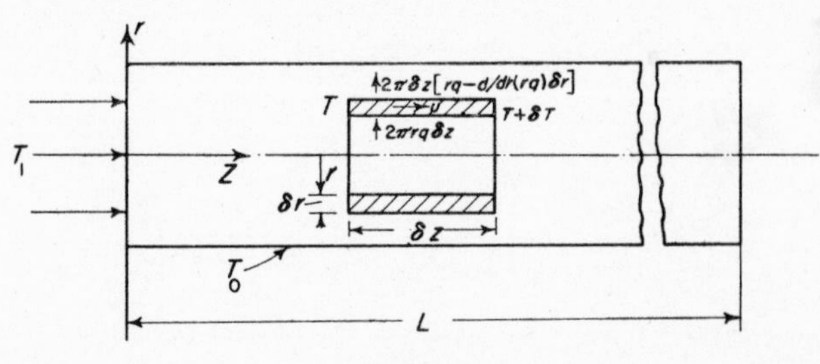

FIG. 57.

A heat balance over a small element of fluid between r and $r + \delta r$ of length δz (Fig. 57), neglecting longitudinal conduction, gives

$$2\pi r \delta r \rho c_p u \delta T = -\ 2\pi\ \partial/\partial r\ (rq)\delta r \delta z \qquad [4.1.1]$$

where q is the heat flux by conduction in a radial direction, i.e.

$$q = -k \frac{\partial T}{\partial r} \qquad [4.1.2]$$

Substituting for q we get

$$\rho c_p u \frac{\partial T}{\partial z} = \frac{1}{r}\frac{\partial}{\partial r}\left(rk\frac{\partial T}{\partial r}\right)$$

which may be written

$$u\frac{\partial T}{\partial z} = \alpha\left[\frac{\partial^2 T}{\partial r^2} + \frac{1}{r}\frac{\partial T}{\partial r}\right] \qquad [4.1.3]$$

where $\alpha = k/\rho c_p$, the thermal diffusivity.

The solution is simplified by introducing the dimensionless temperature coefficient θ where

$$\theta = (T - T_0)/(T_1 - T_0) \qquad [4.1.4]$$

which when substituted into Eqn. [4.1.3] gives

$$u\frac{\partial \theta}{\partial z} = \alpha\left[\frac{\partial^2 \theta}{\partial r^2} + \frac{1}{r}\frac{\partial \theta}{\partial r}\right] \qquad [4.1.5]$$

This is the basic equation which must be solved subject to the boundary conditions,

$$\left.\begin{array}{l}\theta = 0 \text{ at } r = a \text{ for all values of } z > 0 \\ \text{and} \\ \theta = 1 \text{ at } z = 0 \text{ for all values of } r \leqslant a\end{array}\right\} \qquad [4.1.6]$$

The solution depends on the form of u, i.e. on the velocity profile. Three cases can be considered.

(*i*) Piston flow, where u is constant and equal to u_m. This condition applies near the entrance of a pipe and it also represents the limiting condition of 'infinite pseudoplasticity', i,e. when the flow-behaviour index, n, is zero.

(*ii*) Fully developed parabolic velocity profile for a Newtonian fluid. In this case the velocity u in Eqn. [4.1.5] is given by

$$u = 2u_m (1 - r^2/a^2) \qquad [4.1.7]$$

This case also represents the other limit of pseudoplastic flow, i.e. when n becomes unity.

(*iii*) Fully developed velocity profile for a power law fluid when n is neither zero nor unity. The velocity u is now (see Section 3.2) given by

$$u = u_m \frac{(3n + 1)}{(n + 1)} \left[1 - \left(\frac{r}{a}\right)^{\frac{n+1}{n}}\right] \qquad [4.1.8]$$

(a) Solution for piston flow

The equation to be solved is

$$u_m \frac{\partial \theta}{\partial z} = \alpha \left[\frac{\partial^2 \theta}{\partial r^2} + \frac{1}{r}\frac{\partial \theta}{\partial r}\right] \qquad [4.1.9]$$

This can be solved by the method of separation of variables by letting

$$\theta = R(r)\, Z(z)$$

where R is a function of r only and Z is a function of z only.

Eqn. [4.1.9] then reduces to

$$\frac{u_m}{\alpha} \frac{1}{Z} \frac{\mathrm{d}Z}{\mathrm{d}z} = \frac{1}{R} \left[\frac{\mathrm{d}^2 R}{\mathrm{d}r^2} + \frac{1}{r}\frac{\mathrm{d}R}{\mathrm{d}r}\right] \qquad [4.1.10]$$

The left-hand side of Eqn. [4.1.10] is a function of z only and the right-hand side is a function of r only. Then each must be equal to a constant, say $-\beta^2$.

It follows then that

$$\frac{1}{Z}\frac{\mathrm{d}Z}{\mathrm{d}z} = -\frac{\alpha}{u_m}\beta^2$$

for which the solution is

$$Z = \exp\left(-\frac{\alpha}{u_m}\beta^2 z\right) \text{ where } Z = 1 \text{ at } z = 0$$

Further, the radial part of the solution is

$$\frac{\mathrm{d}^2R}{\mathrm{d}r^2} + \frac{1}{r}\frac{\mathrm{d}R}{\mathrm{d}r} + R\beta^2 = 0$$

This is a Bessel equation of zero order for which the solution can be found to be

$$R = 2\sum_{n=1}^{\infty} \frac{1}{\beta_i a}\frac{J_0(\beta_i r)}{J_1(\beta_i a)}$$

where J_0 and J_1 are Bessel functions of the first kind of order 0 and 1 respectively, and the eigen value, β_i, is the ith root of the equation

$$J_0\,(\beta a) = 0$$

The full solution is given by the product

$$\theta = R(r)\,Z(z)$$

Therefore we have

$$\frac{T - T_0}{T_1 - T_0} = \theta = 2\sum_{i=1}^{\infty} \frac{1}{\beta_i a}\frac{J_0(\beta_i r)}{J_1(\beta_i a)} \exp\left(-\frac{\alpha}{u_m}\beta_i^2 z\right) \qquad [4.1.11]$$

This gives the temperature at any point in the pipe as a function of the radial position, r, and the distance down the pipe, z.

This result can be presented in a more convenient form (59). We can define an average heat transfer coefficient, h_{av}, by

$$Q = \pi a^2 u_m \rho c_p (T_2 - T_1) = h_{av}\, 2\pi a\, L \Delta T \qquad [4.1.12]$$

where Q = total rate of heat transfer over length L,
T_2 = average temperature at outlet, i.e. at $z = L$,
ΔT = arithmetic mean temperature difference,
i.e. $\Delta T = T_0 - \frac{1}{2}(T_1 + T_2)$.

Hence
$$h_{av} = \rho u_m c_p \frac{a}{2L}\frac{T_2 - T_1}{\Delta T} \qquad [4.1.13]$$

Further, the average Nusselt number may be defined as

$$\mathrm{Nu} = h_{av}D/k \tag{4.1.14}$$

and by using Eqns. [4.1.11] and [4.1.13] it may be shown that

$$\mathrm{Nu} = \frac{u_m D^2}{2L\alpha}\left(\frac{1-4s}{1+4s}\right)$$

or

$$\mathrm{Nu} = \frac{2}{\pi}\left(\frac{wc_p}{kL}\right)\left(\frac{1-4s}{1+4s}\right) \tag{4.1.15}$$

where S is given by

$$S = \sum_{i=1}^{\infty} \frac{1}{\lambda_i^2} \exp\left(-\pi\lambda_i^2 / \frac{wc_p}{kL}\right) \tag{4.1.16}$$

and λ_i are the roots of $J_0(\lambda) = 0$.

This means that the Nusselt number given by Eqn. [4.1.15] is a function of the Graetz number, wc_p/kL, only. This equation is not convenient for high values of the Graetz number, but in this case it may be approximated by [63]

$$\frac{h_{av}D}{k} = \frac{8}{\pi} + \frac{4}{\pi}\left(\frac{wc_p}{kL}\right)^{\frac{1}{2}}$$

(b) Solution for fully developed velocity profile for Newtonian fluids

On substituting the parabolic velocity profile given in Eqn. [4.1.7] into the basic Eqn. [4.1.5] we get

$$2u_m\left(1-\frac{r^2}{a^2}\right)\frac{\partial\theta}{\partial z} = \alpha\left[\frac{\partial^2\theta}{\partial r^2} + \frac{1}{r}\frac{\partial\theta}{\partial r}\right] \tag{4.1.17}$$

This equation can also be solved by the method of separation of the variables to give a series solution, but the problem is considerably more complicated than that for piston flow.

Leveque [60] has developed an approximate solution to this problem for the case of high rates of mass flow through relatively short tubes by assuming that the velocity gradient is linear near the wall, i.e.

$$u = \alpha'(a-r) \tag{4.1.18}$$

where α' is the velocity gradient at the wall.

The final result may be written

$$\mathrm{Nu} = \frac{h_{av}D}{k} = 1{\cdot}62\left(\frac{\alpha' c_p \rho D^3}{8kL}\right)^{\frac{1}{3}} \qquad [4.1.19]$$

For a Newtonian fluid, α', the velocity gradient at the wall, is $8u_m/D$ and in this case Eqn. [4.1.19] may be written as

$$\frac{h_{av}D}{k} = 1{\cdot}62\left(\frac{4}{\pi}\frac{wc_p}{kL}\right)^{\frac{1}{3}}$$

or

$$\frac{h_{av}D}{k} = 1{\cdot}75\left(\frac{wc_p}{kL}\right)^{\frac{1}{3}} \qquad [4.1.20]$$

where w is the mass flow rate and wc_p/kL is the Graetz number. This equation holds for values of the Graetz number greater than 100.

This result will be used later in the discussion on the laminar flow of non-Newtonian fluids in Section (*d*).

(c) Solution for fully developed velocity profile for a power law fluid

The velocity profile for fully developed laminar flow of a power law fluid is given as Eqn. [4.1.8] and substitution in Eqn. [4.1.5] gives

$$u_m\left(\frac{3n+1}{n+1}\right)\left[1-\left(\frac{r}{a}\right)^{\frac{n+1}{n}}\right]\frac{\partial\theta}{\partial z} = \alpha\left[\frac{\partial^2\theta}{\partial r^2}+\frac{1}{r}\frac{\partial\theta}{\partial r}\right] \qquad [4.1.21]$$

Separating the variables by substituting

$$\theta = R(r)\,Z(z)$$

(as in the case of piston flow) we get the subsidiary equations

$$\frac{1}{Z}\frac{\mathrm{d}Z}{\mathrm{d}z} = -\frac{\alpha}{cu_m}\beta^2 \text{ where } c = \frac{3n+1}{n+1} \qquad [4.1.22]$$

and

$$\frac{\mathrm{d}^2R}{\mathrm{d}r^2}+\frac{1}{r}\frac{\mathrm{d}R}{\mathrm{d}r}+\beta^2\left[1-\left(\frac{r}{a}\right)^{(n+1)/n}\right]R = 0 \qquad [4.1.23]$$

When $n = 0$ these reduce to the equations given in Section (*a*) for piston flow and when $n = 1$ we get the equations given in Section (*b*) for Newtonian flow.

Eqn. [4.1.23] is of the Sturm-Liouville type and Lyche and Bird[61] have derived the complete solutions for the cases $n = 1$, $\frac{1}{2}$ and $\frac{1}{3}$ at low values of the Graetz number. They calculated the eigen values for Eqn. [4.1.23], i.e. values of $\beta_i a$, and tabulated these together with the corresponding coefficients in the series solution for R, for values of $i = 1, 2$ and 3. Corresponding values of the eigen functions R_i are tabulated for values of r/a from 0 to 1 in intervals of 0·1.

In order to indicate the variation in the heat transfer characteristics with the flow behaviour index, n, Lyche and Bird have plotted the reduced temperature θ as a function of r/a at a value of the dimensionless Graetz number wc_p/kL of 5·24 for values of $n = 1, \frac{1}{2}, \frac{1}{3}$ and 0; $n = 1$ is the Newtonian case and $n = 0$ is the case of piston flow which was derived in Section (*a*). This is given in Fig. 58.

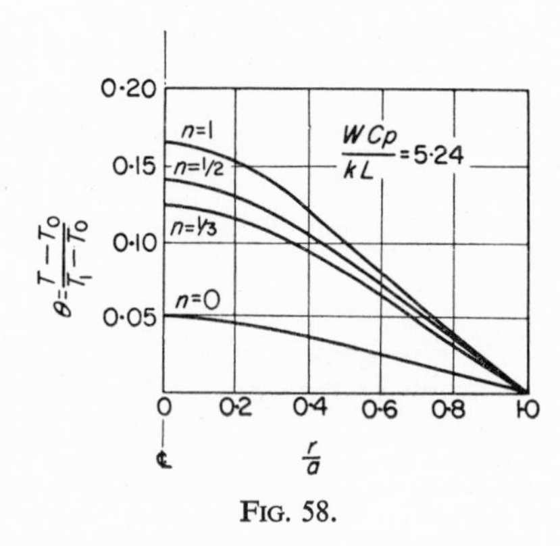

FIG. 58.

(d) Extension of the Leveque approximation to non-Newtonian systems

The Leveque approximation for the average Nusselt number for laminar flow in a pipe has been given as

$$\frac{h_{av}D}{k} = 1{\cdot}62\left(\frac{a' c_p \rho D^3}{8kL}\right)^{\frac{1}{3}} \quad [4.1.19]$$

where a' is the velocity gradient at the wall.

Pigford[62] has rewritten this in the form

$$\frac{h_{av}D}{k} = 1{\cdot}75\ \delta^{\frac{1}{3}}\left(\frac{wc_p}{kL}\right)^{\frac{1}{3}} \quad [4.1.24]$$

where δ is the ratio of the velocity gradient at the wall for the non-Newtonian fluid, a', to that for a Newtonian fluid, $8u_m/D$, i.e. δ is defined as

$$\delta = \frac{a'}{8u_m/D} \quad [4.1.25]$$

In Eqn. [4.1.24] $\delta^{\frac{1}{3}}$ may be regarded as a factor which corrects the Newtonian Eqn. [4.1.20] for changes in heat transfer rate due to the fact the velocity gradient at the wall is different for non-Newtonian fluids.

Pigford has also shown that δ for Bingham plastics* is given by

$$\delta = \frac{1 - \tau_y/\tau_w}{1 - \frac{4}{3}(\tau_y/\tau_w) + \frac{1}{3}(\tau_y/\tau_w)^4} \qquad [4.1.26]$$

and in general, for *all* time-independent fluids by

$$\delta = \frac{(3n' + 1)}{4n'} \qquad [4.1.27]$$

Using this value of δ, Eqn. [4.1.24] applies for high flow rates (with Graetz numbers greater than 100) and values of n' above 0·1. These conditions will normally be encountered in practice so this restriction is not serious. For the rare case of fluids showing extreme pseudoplasticity at low Graetz numbers, Metzner, Vaughn and Houghton (63) have presented an empirical correction in place of $\delta^{\frac{1}{3}}$ which is in the form of an interpolation between the two limiting theoretical solutions for $n' = 0$ and $n' = 1$.

It is worth noting here that the correction for dilatant fluids is never very great, for in the limiting case of 'infinite dilatancy' $n' = \infty$, and Eqn. [4.1.27] gives $\delta = 3/4$. Hence from Eqn. [4.1.24] we get for infinite dilatancy

$$\frac{h_{av}D}{k} = 1{\cdot}59 \left(\frac{wc_p}{kL}\right)^{\frac{1}{3}} \qquad [4.1.28]$$

which is only slightly different from Eqn. [4.1.20] for Newtonian fluids. The correction for pseudoplasticity could be greater as shown by Fig. 59 which shows the curves for a Newtonian fluid (Eqn. [4.1.20]) together with those for infinite dilatancy (Eqn. [4.1.28]) and infinite pseudoplasticity, $n' = 0$, which is identical with the case for piston flow presented in Section (*a*).

Metzner *et al.* (63) have also suggested an empirical correction factor to take into account deviations from theory caused by the distortion of the velocity profile by changes in viscosity due to the radial temperature gradient. This is a generalization of the Sieder-Tate viscosity ratio, $(\mu/\mu_w)^{0.14}$, which is widely used in heat transfer correlations for Newtonian fluids. The denominator in the generalized Reynolds number (Section 3.1.*f*.) is $k'8^{n'-1}$ and this

* A theoretical solution obtained by substituting the appropriate velocity profile for Bingham plastic flow into the basic equation, Eqn. (4.1.5.), has recently been reported by Hirai, E., A.I.Ch.E. Journal 1959 **5** 130.

takes the place of the viscosity in the conventional Reynolds number. Therefore in place of the Sieder-Tate correction Metzner *et al.* have suggested $(m/m_w)^{0.14}$ where $m = k'8^{n'-1}$. m is evaluated at the main bulk temperature and m_w at the wall temperature. The final correlation for the average heat transfer coefficient then becomes

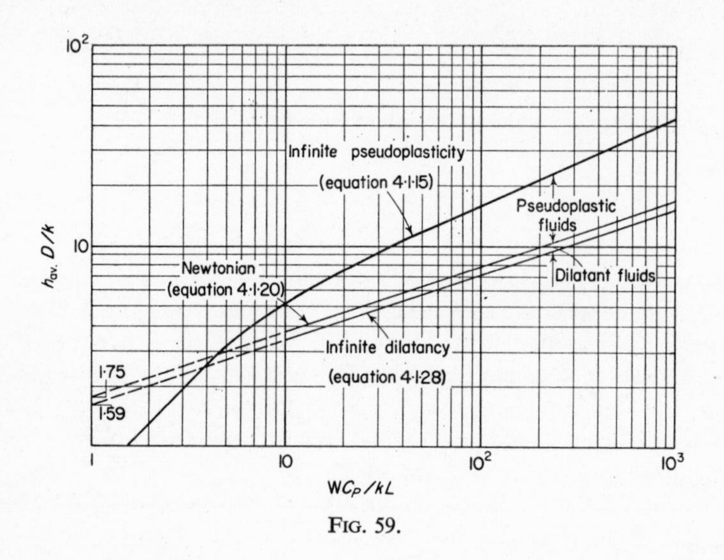

FIG. 59.

$$\frac{h_{av}D}{k} = 1{\cdot}75\ \delta^{\frac{1}{3}} \left(\frac{wc_p}{kL}\right)^{\frac{1}{3}} \left(\frac{m}{m_w}\right)^{0.14} \qquad [4.1.29]$$

This equation was tested experimentally, (63) and found to be successful, over the following range of variables:

n' : 0·18 to 0·70

wc_p/kL : 100 to 2050

Re$'$: 0·65 to 2100

The inclusion of the correction factor $(m/m_w)^{0{\cdot}14}$ in Eqn. [4.1.29] considerably improved the correlation of the experimental results.

(e) Temperature and velocity profiles in non-isothermal flow

In deriving the theoretical relations in the preceding sections it was necessary to assume that the fluid properties were independent of temperature and that frictional heat generation was negligible. These assumptions, which are well suited to conventional fluid flow theory, are more questionable when one is considering the flow of highly viscous non-Newtonian fluids. The viscosity of these fluids often changes markedly with temperature and, because of the high stresses involved in viscous systems, the frictional heat generation is appreciable. Cooling due to expansion can also be significant at the high pressures which are often involved. Variations in thermal diffusivity are also significant for many of these fluids.

Gee and Lyon [64] have taken all these factors into account and they derived the equations which govern the flow of a non-Newtonian fluid through a pipe which may be heated or cooled by heat transfer from the walls during flow. The rheological equation of the fluid was assumed to be of the form

$$\dot{\gamma} = \frac{\tau}{\mu_0}(1 + C\tau^n)$$

When $n = 2$ this is identical with the equation proposed by Rabinowitsch. The effect of temperature on viscosity was allowed for in the usual way, i.e.

$$\mu_0 = Ae^{E/RT}$$

The non-linear partial differential equations which result from this analysis are not amenable to analytical solution. Gee and Lyon used a computer solution to give the steady-state temperature and velocity distributions as a function of radial position and tube length.

The calculations were carried out for a Lucite acrylic resin which was forced through tubes of $\frac{1}{8}$ in. diameter by 4 in. long to $\frac{1}{4}$ in. diameter by 16 in. long. The pressure range was 800–3000 lb/in^2 and the wall temperature 150–250°C. The initial melt temperature was 250°C and here it should be noted that at this temperature Lucite has a Newtonian viscosity of 10^4 poise and this increases by a factor of 2 at 240°C.

The calculations were tested by comparing the predicted and experimental flow rates, i.e. the integrated velocity. The agreement observed was used as evidence of the accuracy of the predicted velocity and temperature profiles (since the local velocities are very sensitive to point temperatures).

4.2 HEAT TRANSFER WITH TURBULENT FLOW IN A PIPE

There is no comprehensive theory of turbulence which allows direct calculation as in the case of laminar flow. In this case dimensional analysis is found to be useful.

(a) Review of Newtonian correlations

For Newtonian fluids we can assume that the local heat flux, q, will depend on a characteristic length of the apparatus, L, the mean temperature difference ΔT, the mean velocity u_m, and on the physical properties of the fluid k, ρ, μ and c_p. Hence we can write

$$q = \phi(\Delta T, k, c_p, \rho, \mu, u_m, L)$$

Applying the principles of dimensional analysis we obtain

$$\frac{hL}{k} = \phi\left(\frac{\rho u_m L}{\mu}, \frac{\mu c_p}{k}\right) \qquad [4.2.1]$$

or

$$\text{Nu} = \phi\ (\text{Re, Pr})$$

The form of this function will depend on the geometry of the system. For fully developed turbulent flow it is found that the function is of the form

$$\text{Nu} = C\ \text{Re}^x\ \text{Pr}^y$$

In this expression h in the Nusselt group is a mean value equal to the average heat flux divided by the logarithmic mean temperature difference. The physical properties of the fluid are evaluated at the mean bulk temperature of the fluid.

C is a shape factor and depends on the geometry of the system. x and y depend on the geometry of the system also. x lies between 0·5 and 0·8 and y between 0·3 and 0·4. For example, for turbulent flow in round pipes

and

$$\left.\begin{aligned} \text{Nu} &= 0{\cdot}023\ \text{Re}^{0\cdot 8}\ \text{Pr}^{0\cdot 4} \text{ for heating} \\ \text{Nu} &= 0{\cdot}023\ \text{Re}^{0\cdot 8}\ \text{Pr}^{0\cdot 3} \text{ for cooling} \end{aligned}\right\} \qquad [4.2.2]$$

where the characteristic length is now the pipe diameter.

The working formulae for other geometrical arrangements, such as flow across banks of tubes as in a heat exchanger shell, are given in the handbooks.

(b) Heat transfer to non-Newtonian fluids

From the point of view of turbulent heat transfer it is convenient to divide non-Newtonian fluids into two classes as follows:

(*i*) fluids which are only slightly non-Newtonian such as dilute suspensions;

(*ii*) fluids which exhibit highly non-Newtonian flow behaviour.

(i) Turbulent heat transfer to slightly non-Newtonian fluids

Several workers, notably Winding *et al.*[65] and Orr and Dalla Valle[66] have investigated the heat transfer characteristics of suspensions which are slightly pseudoplastic (i.e. n in Eqn. [1.2.3] does not differ greatly from unity) in the turbulent region.

These investigations showed that the conventional forms of the Dittus-Boelter equation for forced convection in circular tubes, i.e.

$$\frac{hD}{k} = 0{\cdot}023 \left(\frac{\rho u_m D}{\mu}\right)^{0\cdot 8} \left(\frac{\mu c_p}{k}\right)^{0\cdot 4 \text{ or } 0\cdot 3} \qquad [4.2.3]$$

or the alternative form incorporating the correction of Sieder and Tate, i.e.

$$\frac{hD}{k} = 0{\cdot}023 \left(\frac{\rho u_m D}{\mu}\right)^{0\cdot 8} \left(\frac{\mu c_p}{k}\right)^{0\cdot 33} \left(\frac{\mu}{\mu_w}\right)^{0\cdot 14} \qquad [4.2.4]$$

appear to be applicable provided that sufficient care and thought is given to the measurement of the appropriate fluid properties. Thermal conductivities and viscosities are very difficult properties to determine for suspensions which tend to settle out rapidly. Orr and Dalla Valle have described methods whereby these measurements can be made accurately. The viscosity should be measured in the range of shear stresses to be encountered in the heat exchangers under consideration, but since these correlations are only intended for slightly non-Newtonian systems this last restriction is not serious.

(ii) Turbulent heat transfer to highly non-Newtonian fluids

An application of dimensional analysis to the problem of the turbulent heat transfer to Newtonian fluids has suggested that the relevant dimensionless groups involved are the Reynolds number, $\rho u_m D/\mu$, the Prandtl number, $\mu c_p/k$, and the Nusselt number, hD/k. Metzner *et al.*[63] have suggested that these groups could be generalized to include non-Newtonian systems if the viscosity term in the Reynolds and Prandtl numbers is suitably chosen.

It has been found that the generalized Reynolds number defined by

$$\text{Re}' = D^{n'} u_m^{2-n'} \rho/m \text{ where } m = k' \, 8^{n'-1}$$

is highly successful in correlating pressure drop data in pipe flow problems. Metzner *et al.* suggest that a suitable apparent viscosity for use in the generalized Prandtl number could be found by equating the Newtonian Reynolds number to the generalized one, as follows:

$$\rho u_m D/\mu_a = D^{n'} u_m^{2-n'} \rho/m$$

giving

$$\mu_a = m(u_m/D)^{n'-1}$$

Substituting this into the Prandtl number gives

$$\frac{\mu_a c_p}{k} = \frac{c_p m}{k} (u_m/D)^{n'-1}$$

Also, in place of the Sieder-Tate viscosity ratio correction they propose the generalized factor $(m/m_w)^{0.14}$.

The final correlation suggested then becomes

$$\frac{hD}{k} = 0{\cdot}023\left(\frac{D^{n'}u_m^{2-n'}\rho}{m}\right)^{0{\cdot}8}\left(\frac{mc_p}{k}(u_m/D)^{n'-1}\right)^{0{\cdot}4}\left(\frac{m}{m_w}\right)^{0{\cdot}14} \qquad [4.2.5]$$

This correlation reduces to the generally accepted form, Eqn. [4.2.4], when the fluid is Newtonian, i.e. when $n' = 1$ and $m = \mu$.

Although this approach is an interesting possibility and is useful for predicting the order of magnitude of the turbulent heat transfer coefficient, it should be used with caution, because the experimental verification has not been very extensive up to the present time.

4.3 HEAT TRANSFER AND SKIN FRICTION

Another approach to the evaluation of turbulent heat transfer coefficients is by analogy between heat transfer and skin friction in turbulent flow. The well-known relations for Newtonian systems are the Reynolds and Taylor-Prandtl analogies. These will be briefly reviewed. More details can be found in the standard textbooks, e.g. Kay.[59]

(a) Reynolds analogy

Consider the case of turbulent flow near the wall of a pipe where the time-average, or mean velocity may be expressed as a function only of the distance y from the wall, i.e.

$$u = f(y)$$

The random turbulent fluctuations in velocity will be superimposed on the mean flow and it is convenient to picture this turbulent motion being produced by the movement of lumps of fluid, back and forth across the stream, as illustrated diagrammatically in Fig. 60.

Owing to the transverse velocity gradient, this random movement must involve the transfer of momentum and hence turbulent shearing stresses are set up in the fluid. If there is a transverse temperature gradient heat transfer will take place by the same mechanism.

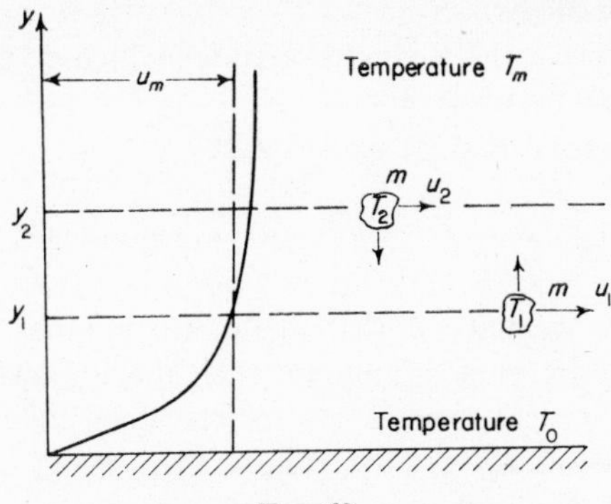

FIG. 60.

Suppose a lump of fluid of mass m moves from a level y_2 to y_1 and that it carries with it the momentum and temperature appropriate to level y_2. To satisfy the condition of continuity another lump of mass m will move from level y_1 to y_2. Hence we have:

$$\textit{net transfer of momentum towards the surface} = m(u_2 - u_1)$$

and

$$\textit{net transfer of heat away from the surface} = mc_p(T_2 - T_1).$$

The turbulent mixing acts in the same way as a shearing stress and the virtual turbulent shearing stress is equal to the increase in momentum generated per unit time. Further, if the lumps move on the average from a region in which the mean velocity is u_m and the mean bulk temperature is T_m up to the surface where the velocity is zero and the temperature is T_0, we have

$$\frac{\textit{rate of transfer of heat}}{\textit{rate of transfer of momentum}} = \frac{q}{\tau_w} = -\frac{c_p(T_m - T_0)}{u_m} = \frac{c_p\Delta T}{u_m}$$

where
$$\Delta T = T_0 - T_m.$$

Therefore
$$\frac{q}{\rho u_m c_p \Delta T} = \frac{\tau_w}{\rho u_m^2} = \tfrac{1}{2}\, c_f$$

or
$$\text{St} = \tfrac{1}{2}\, c_f \qquad [4.3.1]$$

where St, the Stanton number, is defined by $q/\rho u_m\, c_p \Delta T$.

This relation is only approximate since the picture of turbulence on which it is based is over-simplified. In particular it is not correct to assume that the fluid is turbulent right up to the wall. There will be a laminar sublayer.

(b) Taylor-Prandtl analogy

This is an extension of the Reynolds analogy which takes into account the existence of the laminar sub-layer.

By similar reasoning it may be shown that

$$\mathrm{St} = \tfrac{1}{2}\, c_f / [1 + (u_L/u_m)\,(\mathrm{Pr} - 1)] \qquad [4.3.2]$$

where u_L is now the velocity at the edge of the laminar sublayer. This relation reduces to the Reynolds analogy when the Prandtl number is unity.

Various semi-theoretical expressions for the ratio u_L/u_m have been suggested. If the expression due to Hoffmann is used, i.e.

$$u_L/u_m = 1{\cdot}5\ \mathrm{Re}^{-\frac{1}{8}}\ \mathrm{Pr}^{-\frac{1}{6}}$$

we get

$$\mathrm{St} = \tfrac{1}{2}\, c_f / [1 + 1{\cdot}5\ \mathrm{Re}^{-\frac{1}{8}}\ \mathrm{Pr}^{-\frac{1}{6}}\,(\mathrm{Pr} - 1)] \qquad [4.3.3]$$

and this is found to correlate experimental results for heat transfer if the Prandtl number does not differ greatly from unity.

(c) Analogies at high Prandtl numbers and extensions to non-Newtonian systems

Recently, a semi-theoretical analogy between heat, mass and momentum transfer in Newtonian systems at high Prandtl numbers has been presented by Metzner and Friend.(67) This allows for the recent indications that the sub-layer next to the wall is not purely laminar.

The result may be written.

$$\mathrm{St} = \tfrac{1}{2}\, c_f / [1{\cdot}20 + b(\mathrm{Pr} - 1)\,\sqrt{(c_f/2)}] \qquad [4.3.4]$$

The coefficient b has to be determined by experiment. The results were found to be well correlated by

$$b = 11{\cdot}8\ \mathrm{Pr}^{-\frac{1}{3}} \qquad [4.3.5]$$

and using this value of b Eqn. [4.3.4] proved to be satisfactory for Prandtl numbers between 0·46 and 590.

For mass transfer we would have an equation similar to Eqn. [4.3.4] with the Sherwood number, given by k_L/u_m, replacing the Stanton number and the Schmidt number, given by $\mu/\rho\mathscr{D}$, replacing the Prandtl number.

This analogy has now been extended to include the additional complication of non-Newtonian flow properties.(68) An equation, similar to Eqn. [4.3.4], namely

$$\mathrm{St} = \tfrac{1}{2}\, c_f / [1{\cdot}15 + c\sqrt{(c_f/2)}] \qquad [4.3.6]$$

where c is a function of the Prandtl number only, was found to correlate the turbulent heat transfer coefficients in smooth tubes over the following ranges of variables:

Fluids type	*Flow behaviour index n′*	*Prandtl number*	*Reynolds number*
Polymer solutions	0·48–0·87	7·0–90	5,000–37,000
Slurries	0·39–0·92	11–34	7,400–72,000
Newtonian fluids	1·00	1·9–260	5,800–121,000

The range of application of Eqn. [4.3.6] was found to increase as the fluid becomes increasingly pseudoplastic.

CHAPTER 5

MIXING CHARACTERISTICS OF NON-NEWTONIAN FLUIDS

The application of the basic principles of fluid mechanics to the problems of mixing and the design of mixing equipment has only recently been attempted, and at the present time there is no generally accepted method which is capable of predicting quantitatively the performance of mixers even for the comparatively simple Newtonian fluids. A good review of the basic problems has been given by Rushton.[69]

One notable advance in the field of mixing has been the recent work of Hixson, Drew and Knox[70] who have extended the concept of the 'height of a transfer unit', which was developed for steady-state diffusional processes such as absorption and distillation, to the problem of batch operations involving the transfer of material between phases as in a mixing process. They introduced the concept of the 'time of a transfer unit' and this promises to be a profitable line of attack, but much work needs to be done before a generally applicable design procedure can be evolved from it.

Dimensional analysis, a method which has been successfully applied to other problems in fluid mechanics which have proved too complex for rigorous mathematical analysis, can also be used in mixing operations for the prediction of power requirements. The results for Newtonian systems will be briefly reviewed and the extensions to non-Newtonian systems will then be considered.

When one considers the obvious industrial importance of the problem of the agitation of non-Newtonian fluids it is surprising to discover that there is only one published paper, by Metzner and Otto,[71] which attempts to treat this problem in a quantitative manner. There are, however, published results on specialized applications, for example the work of Lee *et al.*[72] on polystyrene solutions and Foresti[73] on suspensions and polymer solutions. Although these data are most valuable for the purpose for which they were intended they do not throw much light on the general problem.

As far as the *goodness of mixing* is concerned there are few published data even for Newtonian systems and, at the present time, the problem has scarcely been attempted quantitatively. Such being the case, the '*equal power rule*' (discussed later) is generally used in default of anything better. Even less information is available for non-Newtonian systems but preliminary qualitative observations [71] appear to indicate that considerably more power is required for the rapid mixing of highly pseudoplastic materials than for Newtonian fluids of similar consistency.

5.1 REVIEW OF MIXING OF NEWTONIAN FLUIDS

(a) Power requirements for mixing of Newtonian fluids

We can assume that the power required for stirring will be a function of the physical properties of the fluid, μ and ρ, the speed of the stirrer, N, the diameter of the stirrer, D, and the gravitational acceleration, g.

Therefore we assume that

$$P = \phi(\mu, \rho, N, D, g)$$

Applying dimensional analysis we should get

$$\frac{P}{\rho N^3 D^5} = \phi\left(\frac{\rho N D^2}{\mu}, \frac{N^2 D}{g}\right) = \phi\,(\text{Re, Fr}) \qquad [5.1.1]$$

This shows that the dimensionless power coefficient $P/\rho N^3 D^5$ depends on both the Reynolds number and the Froude number for a single-phase fluid. If two immiscible liquids are being mixed the interfacial tension would be a factor and we should get

$$\frac{P}{\rho N^3 D^5} = \phi\left(\frac{\rho N D^2}{\mu}, \frac{N^2 D}{g}, \frac{N^2 D^3 \rho}{\sigma}\right) \qquad [5.1.2]$$

where $N^2 D^3\ \rho/\sigma$ is the Weber group, σ being the surface tension.

In a baffled mixer, or in a mixer with an offset stirrer to eliminate swirl, the gravitation effect will be negligible and we can neglect the Froude number in the above functional relations.

For a single-phase fluid it has been found that the functional relation reduces to an equation of the form

$$\frac{P}{\rho N^3 D^5} = C\left(\frac{\rho N D^2}{\mu}\right)^x \left(\frac{N^2 D}{g}\right)^y \qquad [5.1.3]$$

C, x and y depend on the geometry of the system and on the state of flow. For instance x varies from -1 in laminar flow to zero for fully developed turbulence in baffled mixers. In baffled mixers y is also zero.

A typical plot relating the power coefficient to the Reynolds number is shown in Fig. 61. Hence for laminar flow ($\text{Re} < 10$) we can say that the power coefficient is inversely proportional to the Reynolds number, and for fully developed turbulent flow (baffled mixer) when the Reynolds number is greater than about 10^2 the power coefficient is not dependent on the Reynolds number. Between these two cases we have the transition zone which, for mixing, is very wide.

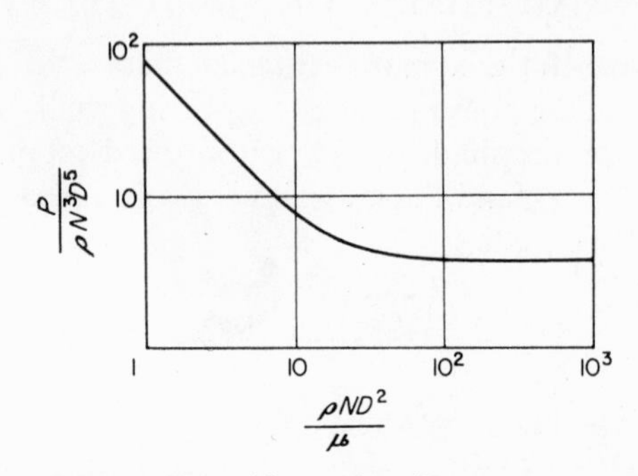

FIG. 61. Power coefficient v. Reynolds number diagram for Newtonian fluids.

(b) Extrapolation for Newtonian fluids

One common method of predicting the power requirements of a proposed large-scale mixer is by the use of models. In unbaffled mixers the power coefficient depends on both the Reynolds number and the Froude number. For true similarity of model and prototype both these groups must be equal. This cannot be achieved if the same fluid is used for the model experiment as will be used in the large-scale prototype. We must use a fluid of lower kinematic viscosity in the model experiment.

For equality of Reynolds numbers

$$(ND^2/v)_L = (ND^2/v)_M$$

Hence
$$v_L/v_M = (N_L/N_M)(D_L/D_M)^2$$

For equality of Froude numbers $(N^2D)_L = (N^2D)_M$

Hence
$$N_L/N_M = (D_M/D_L)^{1/2}$$

Substituting for N_L/N_M we get that

$$v_M/v_L = (D_M/D_L)^{3/2}$$

i.e. v_M must be less than v_L.

The relative speeds will then be given by

$$N_M/N_L = (v_L/v_M)^{1/3}$$

Since the Reynolds and Froude numbers are equal the power coefficients will also be equal and we can now predict the power requirements for the prototype from

$$(P/\rho N^3D^5)_L = (P/\rho N^3D^5)_M$$

or
$$P_L/P_M = (\rho_L/\rho_M)(v_L/v_M)^{7/3}$$

For a baffled mixer the effect of the Froude number is insignificant and equality of Reynolds numbers is all that is required for similarity. Here we could use the same fluid as would be used in the prototype. The condition governing relative speeds and diameters is

$$(ND^2/v)_L = (ND^2/v)_M, \text{ i.e. equality of Re.}$$

Thus $N_M/N_L = (D_L/D_M)^2$ if we use the same fluid.

Hence, using the same fluid, the speed of the model would be excessively high for a useful size ratio. If a quarter-scale model is used it would have to run 16 times as fast. This is excessive and other factors would render the result open to doubt.

However, if the mixer has to work with a very viscous fluid this difficulty could be removed by testing the model with a fluid of low viscosity, for then we should have

$$N_M/N_L = (v_M/v_L)(D_L/D_M)^2$$

and if $v_M \ll v_L$ the speed of the model would not be excessive even for an appreciable scale reduction.

On the whole the use of models is costly in time and money and is best avoided if possible. We can do this by the use of extrapolation formulae, which in effect reduces to the use of an equation such as Eqn. [5.1.3] where C, x and y have been determined for the particular geometrical arrangement under investigation. Various forms of this equation for different types of mixer have been given in the literature and these enable the power requirements to be predicted as a function of the speed and size of the impeller and the physical properties of the fluid.

If we are trying to achieve the same degree of dispersion of, say, a liquid in another liquid or a solid in a liquid the evidence suggests that, in geometrically similar systems using the same materials in the same proportions, the power input per unit volume must be kept the same in both the model and prototype. If the satisfactory degree of dispersion is achieved in the model with a power input P_M then the power input for the prototype to attain the same degree of dispersion would have to be given by

$$P_L/P_M = (D_L/D_M)^3$$

5.2 POWER REQUIREMENTS FOR MIXING OF NON-NEWTONIAN FLUIDS

Of the various types of non-Newtonian fluids, Bingham plastics are the simplest since they can be described by two parameters, the yield stress, τ_y, and the plastic viscosity, μ_p.

Following the procedure used for Newtonian fluids we can say then for Bingham plastics that the power will be given by

$$P = \phi(\rho, \mu_p, \tau_y, N, D, g)$$

Applying dimensional analysis we should get

$$\frac{P}{\rho N^3 D^5} = \phi\left(\frac{\mu_p}{\rho N D^2}, \frac{\tau_y}{\rho N^2 D^2}, \frac{N^2 D}{g}\right)$$

The new group $\tau_y/\rho N^2 D^2$ is seen to have the physical significance of the ratio of the yield stress to the inertia forces. The usual group quoted is the Hedström number He $= \tau_y \rho D^2/\mu_p^2$. This is the product of $\tau_y/\rho N^2 D^2$ and the square of the Reynolds number.

Hence we could write

$$P/\rho N^3 D^5 = \phi\ (\text{Re, He, Fr}) \qquad [5.2.1]$$

Brown and Petsiavas[74] have investigated fluids which are essentially Bingham plastic on these lines.

The technique of dimensional analysis has also been applied to the mixing of other non-Newtonian fluids of the pseudoplastic type by Schultz-Grunow.[75] He studied the power requirements for the agitation of slightly non-Newtonian fluids in the laminar flow region. His correlation is applicable to fluids of the Prandtl type, i.e. those which have a rheological equation of the form

$$\tau = A \sin^{-1} \dot{\gamma}/C \text{ where } A \text{ and } C \text{ are constants.}$$

The correlation is in the form of a logarithmic plot of M/D^3A versus ω/C where M is the torque required at the stirrer and ω is the angular velocity. The range of ω/C is from 0·004 to 2·0 and the range of M/D^3A from 0·02 to 2·0.

Unfortunately this correlation is of limited usefulness because the impellers studied by Schultz-Grunow were of little practical interest (simple crosses of 135–325 mm) and the flow was in the strictly laminar regime where little mixing of non-Newtonian fluids appears to take place.

The only work which appears to offer a useful design procedure for highly non-Newtonian fluids of the pseudoplastic type is that of Metzner and Otto.[71] They emphasized that since the apparent viscosity may vary by as much as a thousand-fold depending on the shear rate, any one value of the apparent viscosity is unlikely to give a good correlation when the range of shear rates which exist in the mixer is appreciably different from the shear rate at which the viscosity was measured.

The basis of Metzner and Otto's method is the determination of the range of shear rate in the mixer and its relation to the variables of the system. In order to do this they defined an apparent viscosity as the viscosity of a Newtonian fluid which requires the same power input under the same conditions of size and speed (in the laminar region). The next stage is to relate this to the other variables of the system. They assumed that the fluid motion in the region of the impeller could be characterized by an average shear rate which was proportional to the rotational speed of the stirrer, i.e.

$$(\dot{\gamma})\ \text{average} \propto N = KN$$

This assumption is reasonable since it can be shown theoretically to be true for a cylinder rotating in an infinite fluid. The value of K has to be found indirectly and the procedure adopted by Metzner and Otto is as follows.

(1) Find μ_a at a given N from the measured power coefficient and the appropriate correlation between power coefficient and Reynolds number used for Newtonian fluids, i.e. measure $P/\rho N^3 D^5$ for the non-Newtonian system, and find, $\rho N D^2/\mu_a$ from the correlation at this value of the power coefficient. This gives μ_a, knowing $\rho N D^2$ for the system, if μ_a is defined as above, i.e. the viscosity of a Newtonian fluid which gives the same power coefficient in the same apparatus at the same speed.

(2) Find the shear rate, $\dot{\gamma}$, which gives this value of μ_a from the flow curve of the non-Newtonian fluid.

(3) Find K by substituting the values of $\dot{\gamma}$ and the speed, N, in the equation

$$(\dot{\gamma})_{av} = KN$$

(4) Repeat for different fluids at different speeds using various sizes of apparatus to determine the best value of K. All the experiments must be confined to the laminar flow region, of course, because outside this region viscosity is not an important factor.

The value of K so obtained by Metzner and Otto was 13 where N is in r.p.m. and $(\dot{\gamma})_{av}$ in min^{-1}.

Metzner and Otto then plotted the results of their experiments (which extended well into the turbulent region) in terms of power coefficient against Reynolds number. In calculating the Reynolds number μ_a is found from the flow curve of the material at a shear rate given by $13N$. This correlation is given in Fig. 62 and compared with the conventional correlation for Newtonian fluids. The two are identical in the laminar flow region, of course, by virtue of the definition of μ_a.

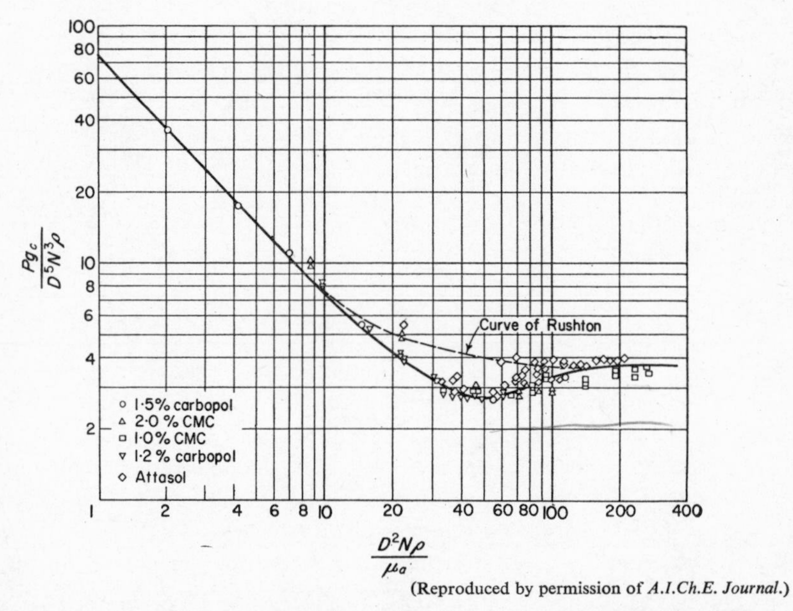

(Reproduced by permission of *A.I.Ch.E. Journal.*)

FIG. 62. Power coefficient v. Reynolds number diagram for non-Newtonian fluids.

The main difference between the two curves is that the laminar region for non-Newtonian pseudoplastic fluids extends further than that for Newtonian fluids. This is because μ_a increases with increasing distance from the impeller (regions of lower shear rate). This tends to depress eddies and delay the onset of turbulence.

Calderbank and Moo-Young[76] have repeated and extended this work for a wide variety of sizes and types of impeller. They found that for Bingham and pseudoplastic fluids the value of K was 10 rather than 13 as suggested by Metzner and Otto. For dilatant fluids the relative size of the impeller and mixing vessel has an influence on the shear rate and therefore on the effective value of K. Calderbank and Moo-Young found in their work that for these fluids K could be represented by $12{\cdot}8\,(D_1/D_T)^{\frac{1}{2}}$ where D_1 and D_T are the diameters of the impeller and tank respectively.

To sum up this method suppose the type, size and speed of the turbine mixer have been fixed (say the speed has to be determined from the conditions of shear which experience suggests is necessary to give the required efficiency of mixing). The procedure to predict the power requirements would then be as follows:

(1) From N find $(\dot{\gamma})_{av} = 13\,N$.
(2) From $(\dot{\gamma})_{av}$ find μ_a from the flow curve.
(3) Calculate $\rho N D^2/\mu_a$ and hence the power coefficient from the curve in Figure 62.

Two points must be remembered here. In the first place Metzner and Otto's curve applies only to flat-bladed turbines. Since the curves for Newtonian and non-Newtonian fluids appear to agree quite closely it would seem permissible to use the Newtonian correlation applicable to other types of mixer as a first estimate. This would be on the safe side and lead to an over-estimate since the Newtonian curve will lie above the curve for the pseudoplastic fluid where it does not coincide with it. The second point is that these fluids are rather unknown quantities and it would be dangerous to extrapolate too far.

CHAPTER 6

VISCOMETRIC MEASUREMENTS AND APPARATUS

The accurate measurement of fluid properties is of course the essential prerequisite to progress in the development of engineering design procedures. The basic principles underlying the experimental characterization of the various types of non-Newtonian fluids have been discussed in Chapter 2. In the present chapter further details of the apparatus and viscometric techniques used will be considered.

Only the two basic instruments will be discussed. These are the capillary and rotational viscometers. Many other empirical methods of measurement have been suggested but the analysis of data from these is often impossible. The capillary and rotational viscometers are both capable of providing reliable quantitative rheological data and they should be regarded as complementary for, as shown in Chapter 2, the data from each type are easily inter-related and compared. Each has its inherent advantage, however. Capillary tube viscometers are obviously to be preferred when the data are to be used for pipe flow problems, and rotational instruments—which subject the material under test to a precise and uniform rate of shear—have definite advantages in the analysis of complex systems, such as those exhibiting thixotropy.

6.1 CAPILLARY TUBE VISCOMETERS

Not many commercial instruments are available which possess the required characteristics and since the construction of capillary tube viscometers is relatively simple it is customary to build these instruments in the laboratory workshop. The essential features of this type of apparatus are merely equipment for the control and measurement of the pressure difference across a capillary tube of known dimensions together with a means of determining of the rate of flow through the tube.

On the basis of considerable experience, workers at the University of Delaware have developed such an instrument and this has recently been described by Dodge.[45] It is shown diagrammatically in Fig. 63. It consists of a heavy brass chamber of about 1 capacity which is sealed at the top and bottom by stainless steel plugs. The precision-bore stainless steel capillary tube (the diameter having been determined by calibration with a Newtonian liquid of known viscosity) is screwed into the bottom plug. Nitrogen at a carefully controlled constant pressure is admitted through the top plug. The rate of flow is measured by collecting a sample over a measured time and weighing.

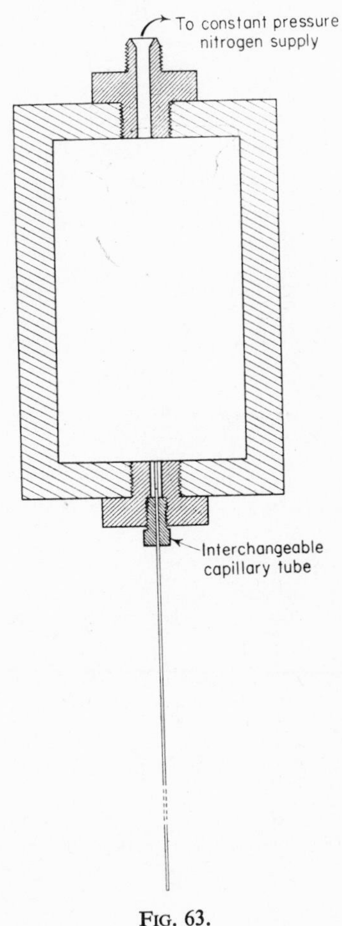

FIG. 63.

One of the main difficulties in capillary tube viscometry is in the correction of the measured overall pressure drop. Three corrections are required, as follows:

(1) for the head of liquid over the tube;
(2) for kinetic energy effects;
(3) for entrance losses.

The first correction is straightforward but some doubt exists about the other two. They can be estimated experimentally by repeating experiments in tubes of different lengths and extrapolating the overall pressure drop to zero length, or by calibration with a Newtonian fluid of known viscosity and preferably of consistency and density similar to those of the fluid under test. The other method is to calculate these effects but, as pointed out in Section 3.5(c), there is some confusion on the question of contraction losses in laminar flow of non-Newtonian fluids. Dodge[45] suggests that the correction for entrance and kinetic energy effects is roughly equal to $1{\cdot}5\rho\, u_m^2/g$ and there is some theoretical justification for this figure. The best way out of the difficulty is to design and operate the viscometer so that the correction is small. This involves the use of a tube with a high ratio of length to diameter.

6.2 ROTATIONAL INSTRUMENTS

Three types of commercially available viscometer will be considered in this section. The basic principles of the three types are:

(*a*) a rotating bob in an infinite fluid;
(*b*) a cone and plate viscometer;
(c) a cone and plate viscometer with provision for the measurement of normal stresses for use with viscoelastic fluids.

(a) The Brookfield Synchro-lectric viscometer

One of the simplest types of instrument which is capable of providing accurate rheological data is that consisting of a cylindrical bob rotating in an infinite fluid. It has been shown in Chapter 2 that the shear stress and rate of shear can be determined at the same point in the fluid, i.e. at the surface of the bob, and by measuring the resisting torque at various speeds the flow curve of the material can be determined directly.

The Brookfield Synchro-lectric viscometer (manufactured by Brookfield Engineering Laboratories, Stoughton, Mass., U.S.A.) is designed on this principle. A photograph of the instrument is given in Fig. 64, and the working parts are illustrated in Fig. 65.

For empirical measurements, or measurements on simple Newtonian liquids, the rotating member is usually in the form of a disk as shown in Fig. 64; to allow quantitative analysis of the data for more complex systems, however, a set of cylindrical spindles can be used. A choice of eight speeds is normally provided, covering a 200 : 1 range with a maximum speed of 600, 100 or 60 r.p.m., depending on the type of instrument. The torque scale is linear, having 100 subdivisions, and the maximum torque can vary between about 700 and 60,000 dyne–cm for different types of instrument.

FIG. 64. Brookfield Synchro-lectric viscometer.

facing p. 122

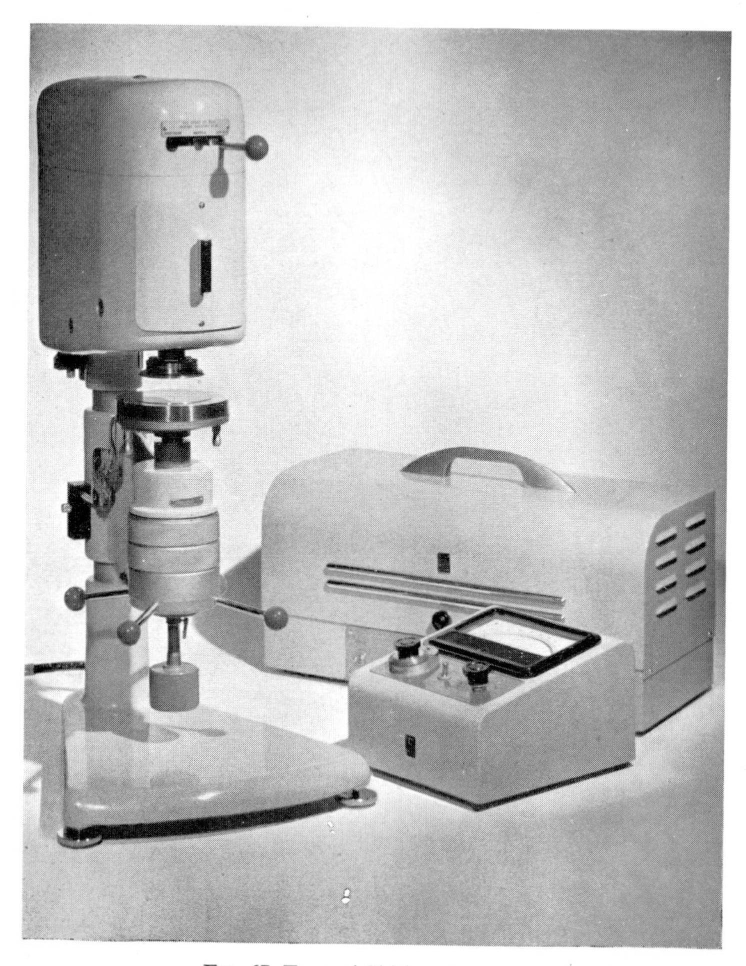

FIG. 67. Ferranti-Shirley viscometer.

An instrument of this type, the Brookfield Viscometran, has been developed for automatic control of viscosity as a process variable.

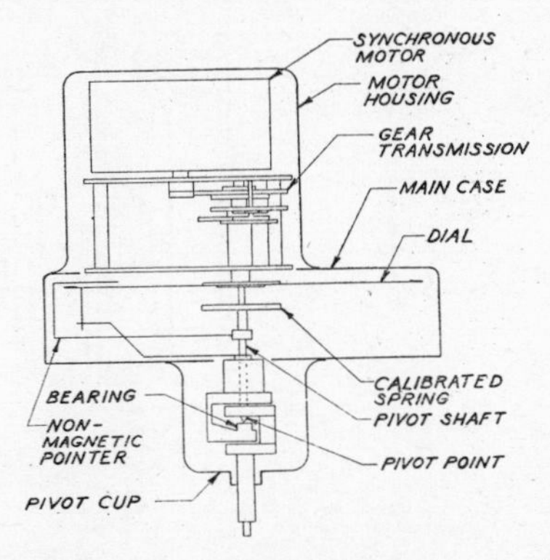

FIG. 65. Brookfield Synchro-lectric viscometer—sectional view.

(b) The Ferranti-Shirley cone and plate viscometer

The advantage of the cone and plate viscometer is that the material is subjected to a uniform rate of shear. This cannot be accomplished with a capillary viscometer and is difficult to achieve with the coaxial cylinder type, for the gap must be very small and this leads to difficulties in construction and causes complications in filling and cleaning.

The Ferranti-Shirley viscometer has been developed by Ferranti Ltd., of Manchester, England, from an original instrument used by the British Cotton Industry Research Association. A sectional diagram of the instrument is given in Fig. 66, and a general view in Fig. 67.

The fluid to be examined (less than 0·5 ml in volume) is sheared between a stationary flat plate and a slightly conical rotating disc which is driven at controlled speeds, ranging from 1 to 1000 r.p.m., by an electronically controlled d.c. servomotor to give infinitely variable speed. Since the speed is directly proportional to the rate of shear, infinitely variable rate of shear over a wide range can be attained. The viscous traction on the cone exerts a torque on a precision electro-mechanical torque dynamometer and this is indicated on an electrical instrument of variable sensitivity. For Newtonian fluids the scale can be calibrated directly in poise; for non-Newtonian fluids the scale readings are directly proportional to apparent viscosity and a conversion factor is employed. The flow curve is then easily constructed.

Thermocouples embedded in the plate (in direct contact with the fluid) enable temperature measurements to be obtained. The plate is also provided with a water jacket through which water at a constant temperature may be circulated.

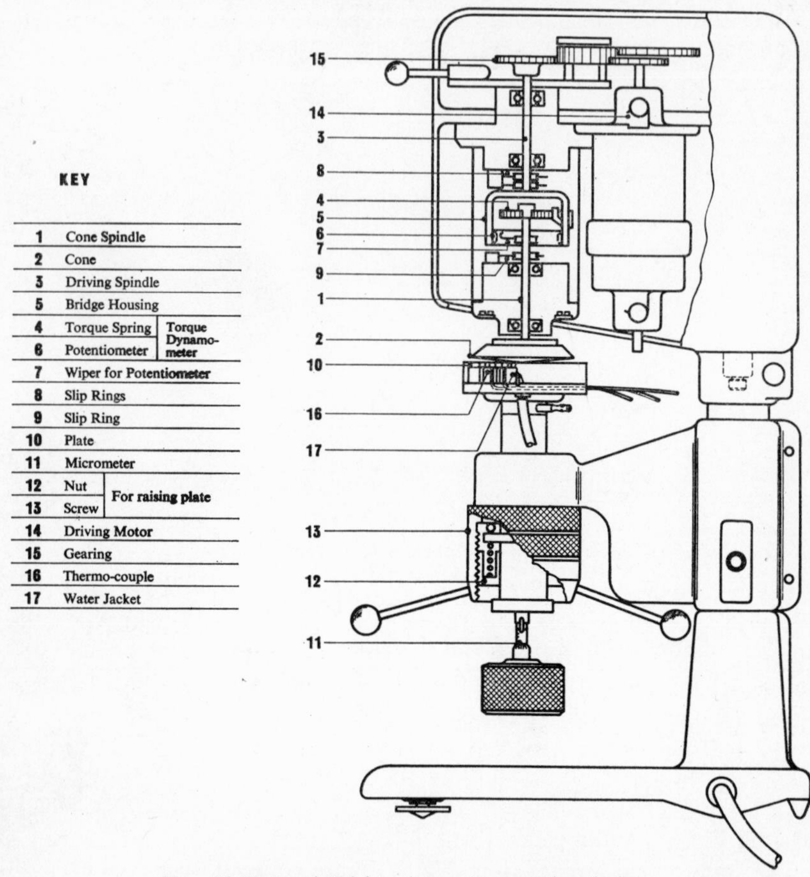

FIG. 66. Ferranti-Shirley viscometer—sectional view.

The instrument is provided with a range of cones and the standard ranges are shown in the following Table.

	Maximum range (1 r.p.m. 500 scale divisions)			Minimum range (1,000 r.p.m. 100 scale divisions)		
	Large cone	Medium cone	Small cone	Large cone	Medium cone	Small cone
Viscosity (poise)	0–707	0–3,870	0–31,555	0–0·14	0–0·75	0–6·3
Shear stress (dyne/cm²)	0–12,725	0–69,660	0–568,000	0–2,545	0–13,532	0–113,600
Rate of shear (sec⁻¹)	18	18	18	18,000	18,000	18,000

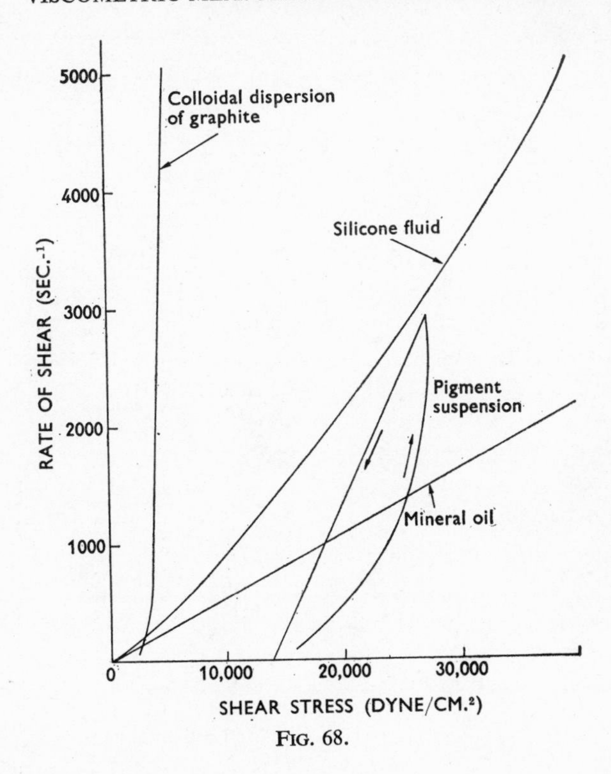
5000
4000
3000
2000
1000
0
10,000
20,000
30,000
RATE OF SHEAR (SEC.$^{-1}$)
SHEAR STRESS (DYNE/CM.^{2})
Colloidal dispersion of graphite
Silicone fluid
Pigment suspension
Mineral oil

FIG. 68.

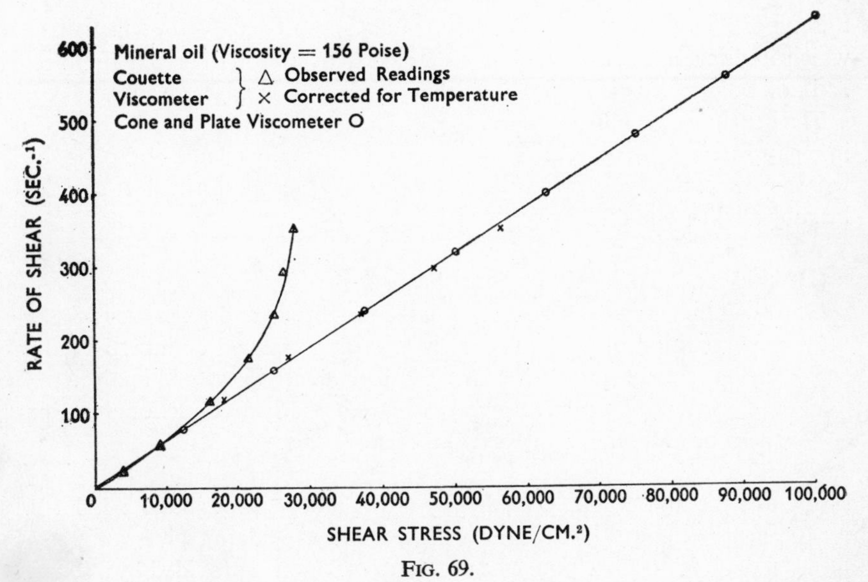
Mineral oil (Viscosity = 156 Poise)
Couette Viscometer } △ Observed Readings
× Corrected for Temperature
Cone and Plate Viscometer O
600
500
400
300
200
100
0
10,000
20,000
30,000
40,000
50,000
60,000
70,000
80,000
90,000
100,000
RATE OF SHEAR (SEC.$^{-1}$)
SHEAR STRESS (DYNE/CM.^{2})

FIG. 69.

Typical flow curves obtained with this instrument are given in Fig. 68. Fig. 69 shows a comparison between the Ferranti-Shirley cone and plate viscometer and a conventional coaxial viscometer with an annular gap width of 0·6 mm and a water-cooled outer cylinder. The coaxial viscometer requires an appreciable temperature correction above 70 sec^{-1} to compensate for the heat produced by stress in the fluid. On the other hand the very thin layer of fluid and the efficient heat transfer of the cone and plate system allow measurements to be made at high rates of shear without the need for compensation for temperature errors. The cone and plate system is therefore capable of discriminating between the effects of shear and temperature on the decrease of apparent viscosity, a problem which has often caused ambiguity in the interpretation of data obtained with coaxial rotational viscometers.

(c) The Roberts-Weissenberg Rheogoniometer

Conventional rotational viscometers such as the Couette or cone and plate types cannot be used to characterize viscoelastic fluids since they are only capable of measuring the tangential component of stress along the streamlines when the fluid is in laminar shear. Viscoelastic fluids give rise to a normal stress, as discussed in Chapter 2, and measurement of this stress allows these fluids to be characterized.

Weissenberg[77] has formulated the principles of an apparatus which would enable measurments to be made of the components of stress and strain round the full solid angle in space at every point in the flowing material. Russell[78] has reported a preliminary development of an instrument approximating to this type. Roberts[79] has since described an improved type of instrument which is discussed below. It is manufactured by Farol Research Engineers Ltd., of Bognor Regis, England.

The instrument is illustrated in Fig. 70. The material under test is contained in a gap between two platens of 3 in. diameter, usually a cone and plate, but whose form can be varied to suit the experiment to be performed. It can then be subjected to one of four types of laminar shearing action as follows:

(1) unidirectional steady laminar shear;
(2) oscillatory laminar shear of variable amplitude and frequency;
(3) superposition of (2) on (1) so that the material can be investigated in various steady states of shear by independent vibrational analysis;
(4) acceleration at a predetermined rate to a steady unidirectional laminar shear or a deceleration from this state.

Motion of the first type is provided by a d.c. motor through a 1 : 1, 1 : 10, 1 : 100 gearbox and this, together with a voltage regulator, gives a range of speeds from 0·05 to 150 r.p.m. The oscillatory motion is provided by a second motor and gearbox which operates a variable-throw cam mechanism M to give a frequency range of 1 to 300 c/min. The amplitude is varied by adjusting the micrometer thimble B', and may range from 0·001 to 0·050 radians. Both drives are combined to give type (3) motion. The rotational motion is imparted by turning the worm W, via the wheels S, whilst the oscillatory motion results from an axial movement of the worm.

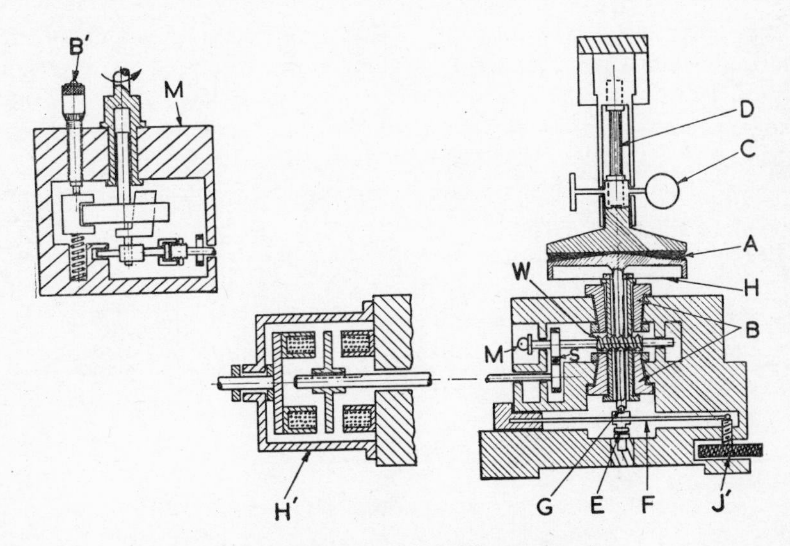

FIG. 70. Roberts-Weissenberg Rheogoniometer.

The normal tangential stresses are measured by recording the displacement of the condenser gauge C, against the torsion bar D. In order to measure normal stresses the head shown in Fig. 31 is used. This consists of series of tubes inserted in a glass plate, and the hydrostatic head of liquid in each tube is taken as a measure of the stress in the axial direction at that point. Some materials take an excessively long time to reach equilibrium in these tubes and a new type of head is now being developed in which the normal stresses at a series of points on its surface can be measured with pressure-actuated condenser gauges.

The total normal force can be measured by transmitting it through a central rod, via the diaphragm H to the spring condenser gauge E, F, where it can be recorded or, if static, measured by a null method using the micrometer screw, J'.

APPENDIX 1

EFFECTIVE SLIP NEAR A SOLID BOUNDARY

It has been mentioned in Chapter 2 that some non-Newtonian fluids show anomalous behaviour near a solid boundary, due to a preferred orientation of particles or molecules caused by the presence of the wall. This effect can be shown with a suspension of paper pulp. When this material flows through a transparent pipe a clear water annulus is observed adjacent to the wall.

The fluid in this region has a lower viscosity and this gives rise to an effective velocity of slip at the wall.

Hence in the region close to the wall (say within a normal distance ϵ) the rate of shear will not be a function of the shear stress alone. If the shear stress near the wall $y = 0$ is τ, the rate of shear will differ from the value $f(\tau)$ which it would have remote from a boundary by an amount which depends on y, the normal distance from the wall,

i.e.
$$\mathrm{d}u/\mathrm{d}y = f(\tau) + g(\tau, y)$$
where
$$g(\tau, y) = 0 \text{ when } y > \epsilon.$$

The velocity u just outside the region of anomalous flow is then

$$u = f(\tau)y + \int_0^\epsilon g(\tau, y)\,\mathrm{d}y$$

In other words

$$u - s(\tau) = f(\tau)y$$

where

$$s(\tau) = \int_0^\epsilon g(\tau, y)\,\mathrm{d}y \qquad [\text{A.1.1}]$$

$s(\tau)$ is the value of u when y is zero, and is the effective velocity of slip at the wall; it is dependent on the local shear stress, τ. This approach has been presented by Oldroyd.[17]

If the boundary condition $u = s(\tau_w)$ at $r = a$ is substituted for the condition of zero velocity at the wall in the derivation of the generalized relation for flow in a pipe (Section 3.1.) we get

$$\frac{Q}{\pi a^3} = \frac{s(\tau_w)}{a} + \frac{1}{\tau_w^{\,3}} \int_0^{\tau_w} \tau^2 f(\tau)\, d\tau \qquad \text{[A.1.2]}$$

This can be written

$$\frac{Q}{\pi a^3 \tau_w} = \frac{\zeta(\tau_w)}{a} + \phi(\tau_w) \qquad \text{[A.1.3]}$$

where

$$\zeta(\tau_w) = s(\tau_w)/\tau_w$$

and

$$\phi(\tau_w) = \frac{1}{\tau_w^{\,4}} \int^{\tau_w} \tau^2 f(\tau) d\tau$$

$\zeta(\tau_w)$ has been called by Oldroyd[80] the *effective slip coefficient*. It is the effective velocity of slip per unit shear stress at the wall.

$\zeta(\tau_w)$ can be evaluated from pipe flow experiments. For a range of pipes of different radius, a, plot $Q/\pi a^3 \tau_w$ against τ_w. If there is slip at the wall the curves will be distinct for each value of a. Then from these curves at a selected value of τ_w determine $Q/\pi a^3 \tau_w$ as a function of $1/a$. When plotted, these points should give a straight line of slope $\zeta(\tau_w)$ and intercept $\phi(\tau_w)$ as seen from Eqn. [A.1.3]. By repeating this at different values of τ_w, $\zeta(\tau_w)$ can be evaluated as a function of τ_w.

Toms[81] has applied this technique to polymer solutions.

APPENDIX 2

BASIC EQUATIONS FOR COAXIAL CYLINDER VISCOMETER

Consider a section through the viscometer, Fig. 1, of unit height in the region of uniform flow. The measured couple per unit height is G.

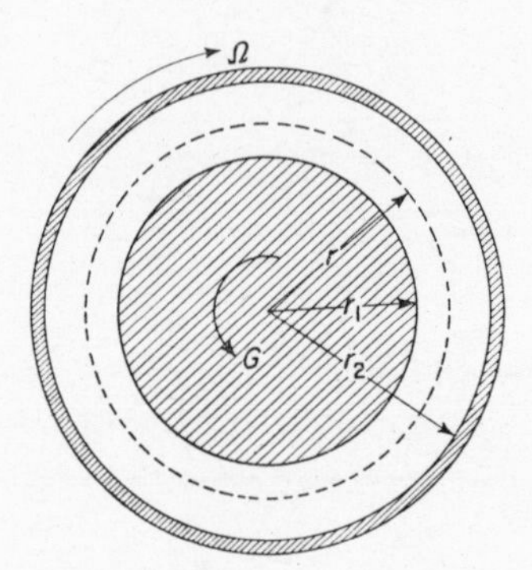

FIG. 1.

We shall consider fluids which are not time-dependent and which can be described by the simple rheological equation

$$\dot{\gamma} = f(\tau)$$

The shear stress $\tau(r)$ at radius r is given by

$$\tau(r) = G/2\pi r^2$$

and the rate of shear by

$$\dot{\gamma} = r\frac{\mathrm{d}\Omega}{\mathrm{d}r} = r\frac{\mathrm{d}}{\mathrm{d}r}\left(\frac{u}{r}\right)$$

Therefore

$$r\frac{\mathrm{d}}{\mathrm{d}r}\left(\frac{u}{r}\right) = f\left(\frac{G}{2\pi r^2}\right) \qquad \text{[A.2.1]}$$

The solution of this differential equation depends on the boundary conditions at the surfaces of the two cylinders. If there is no slipping at the cylinders these become

$$u = 0 \text{ at } r = r_1$$

$$\text{and } u = r_2\Omega \text{ at } r = r_2$$

Hence, integrating Eqn. [A.2.1], we get the velocity distribution as

$$\frac{u(r)}{r} = \int_{r_1}^{r} f\left(\frac{G}{2\pi r^2}\right)\frac{\mathrm{d}r}{r} \qquad [A.2.2]$$

and further when $r = r_2$ we get

$$\Omega = \frac{u(r_2)}{r_2} = \int_{r_1}^{r_2} f\left(\frac{G}{2\pi r^2}\right)\frac{\mathrm{d}r}{r} \qquad [A.2.3]$$

This gives the relation between Ω and G in terms of the function f. Substituting $\tau(r) = G/2\pi r^2$ and $\mathrm{d}r/r = -\,\mathrm{d}\tau/2\tau$ we get

$$\Omega = \int_{\tau_1}^{\tau_2} \frac{f(\tau)\mathrm{d}\tau}{2\tau} \qquad [A.2.4]$$

The solution of these equations depends on the form of the function f. They will now be solved for the case of a Newtonian liquid, a Bingham plastic and a power law fluid.

(a) Newtonian liquid

In the case of a Newtonian liquid of viscosity μ we have

$$f(\tau) = \tau/\mu \text{ where } \tau = G/2\pi r^2.$$

Hence from Eqn. [A.2.2]

$$\frac{u(r)}{r} = \int_{r_1}^{r} \frac{G}{2\pi r^2 \mu}\frac{\mathrm{d}r}{r}$$

i.e.

$$u(r) = -\frac{Gr}{2\pi\mu}\left[\frac{1}{2r^2}\right]_{r_1}^{r} = \frac{G}{4\pi\mu}\left[\frac{r}{r_1^2} - \frac{1}{r}\right] \qquad [A.2.5]$$

Also from Eqn. [A.2.4]

$$\Omega = \int_{\tau_2}^{\tau_1} \frac{d\tau}{2\mu} = (\tau_1 - \tau_2)/2\mu$$

and since $\tau_1 = G/2\pi r_1^2$ and $\tau_2 = G/2\pi r_2^2$ we get

$$\mu = (r_2^2 - r_1^2)G/4\pi r_1^2 r_2^2\,\Omega \qquad \text{[A.2.6]}$$

This equation shows that the viscosity of a Newtonian fluid can be found in principle by measuring G and Ω simultaneously in *one* experiment.

Substituting for G from Eqn. [A.2.6] into Eqn. [A.2.5] gives the velocity distribution in terms of Ω as

$$u(r) = \Omega\,(r - r_1^2/r)/(1 - r_1^2/r_2^2) \qquad \text{[A.2.7]}$$

(b) Bingham plastic

A qualitative picture of the flow pattern when a Bingham plastic is contained in a coaxial cylinder viscometer has been given in Section 3.1(*b*). To find the general relations between Ω and G we proceed as follows. The rheological equation is

$$\tau - \tau_y = \mu_p \dot{\gamma}\ ;\ \tau > \tau_y$$

i.e.

$$f(\tau) = (\tau - \tau_y)/\mu_p$$

Therefore, since in general

$$u(r) = r \int_{\tau(r)}^{\tau_1} f(\tau)\mathrm{d}\tau/2\tau$$

we have for a Bingham plastic

$$u(r) = r \int_{\tau(r)}^{\tau_1} \frac{\tau - \tau_y}{\mu_p}\frac{\mathrm{d}\tau}{2\tau}\ ;\ \tau > \tau_y$$

Thus $$u(r) = \frac{r}{2\mu_p}\Big[\tau - \tau_y \ln \tau\Big]_{\tau(r)}^{\tau_1} = \frac{r}{2\mu_p}\left[\tau_1 - \tau(r) + \tau_y \ln \frac{\tau(r)}{\tau_1}\right]$$

which on substitution of $\tau = G/2\pi r^2$ gives finally

$$u(r) = \frac{G}{4\pi\mu_p}\left[\frac{r}{r_1^2} - \frac{1}{r}\right] - \frac{\tau_y r}{\mu_p}\ln\frac{r}{r_1} \qquad \text{[A.2.8]}$$

This equation holds so long as the local shearing stress, τ, is greater than the yield stress, τ_y, i.e. in the region $r_1 \leqslant r \leqslant r_y$ where r_y is given by

$$\tau_y = G/2\pi r_y^2 \qquad \text{[A.2.9]}$$

If $r > r_y$ we have

$$u(r) = ru(r_y)/r_y;$$

i.e. particles in the fluid have the same angular velocity as they have at radius r_y and the material rotates as a rigid body with the outer cylinder.

The relation between Ω and G depends on whether r_2 is greater or less than r_y. If $r_2 > r_y$, so that some material near the outer cylinder is not sheared, we have that $\Omega = u(r_y)/r_y$. Therefore from Eqn. [A.2.8]

$$\Omega = \frac{G}{4\pi\mu_p}\left[\frac{1}{r_1^2} - \frac{1}{r_y^2}\right] - \frac{\tau_y}{\mu_p}\ln\frac{r_y}{r_1}$$

Substituting for r_y from Eqn. [A.2.9] this becomes

$$\Omega = \frac{G}{4\pi\mu_p r_1^2} - \frac{\tau_y}{2\mu_p}\left[1 + \ln\frac{G}{2\pi r_1^2\tau_y}\right] \qquad \text{[A.2.10]}$$

and the condition is that $2\pi r_1^2\tau_y \leqslant G < 2\pi r_2^2\tau_y$.

If $r_2 < r_y$ so that all the material is sheared

$$\Omega = \frac{u(r_2)}{r_2} = \frac{G}{4\pi\mu_p}\left(\frac{1}{r_1^2} - \frac{1}{r_2^2}\right) - \frac{\tau_y}{\mu_p}\ln\frac{r_2}{r_1} \qquad \text{[A.2.11]}$$

and the condition for this is that $2\pi r_2^2\tau_y \leqslant G$.

If $2\pi r_1^2\tau_y > G$, $\Omega = 0$; i.e., the cylinder will not move.

The material will be sheared throughout when $r_2 = r_y$ in Eqn. [A.2.10] i.e. when the speed reaches a value

$$\Omega = \frac{\tau_y}{\mu_p}\left[\frac{r_2^2}{2r_1^2} - \tfrac{1}{2} - \ln\frac{r_2}{r_1}\right]$$

In Bingham plastic flow in a pipe the solid plug never disappears at a finite rate of throughput.

(c) Power law fluid

We now have that

$$\tau = k(\dot{\gamma})^n$$

$$\text{or } \dot{\gamma} = f(\tau) = (\tau/k)^{1/n}$$

Therefore from Eqn. [A.2.4] we obtain

$$\Omega = \int_{\tau_2}^{\tau_1} \left(\frac{\tau}{k}\right)^{1/n} \frac{d\tau}{2\tau}$$

i.e.

$$\Omega = \frac{n}{2k^{1/n}} \left[\tau_1^{1/n} - \tau_2^{1/n}\right]$$

or

$$\Omega = \frac{n}{2k^{1/n}} \left[\frac{G}{2\pi}\right]^{1/n} \left[\left(\frac{1}{r_1^2}\right)^{1/n} - \left(\frac{1}{r_2^2}\right)^{1/n}\right] \quad \text{[A.2.12]}$$

This gives a relation between Ω and G which depends on n and k, but it is not practicable to find n and k directly.

The variation in rate of shear with radius can be determined for a power law fluid since

$$\dot{\gamma}(r) = \left(\frac{G}{2\pi r^2 k}\right)^{1/n}$$

and substituting for G from Eqn. [A.2.12] we obtain

$$\dot{\gamma}(r) = \frac{2\Omega}{n\left[\left(\frac{1}{r_1}\right)^{2/n} - \left(\frac{1}{r_2}\right)^{2/n}\right] r^{2/n}} \quad \text{[A.2.13]}$$

This shows that the variation in rate of shear across the gap depends on n, and even if the variation is not significant for a Newtonian fluid it could be appreciable for a highly pseudoplastic fluid in the same apparatus.

The apparent viscosity as a function of the radius follows from

$$\mu_a(r) = \tau(r)/\dot{\gamma}(r)$$

$\tau(r) = G/2\pi r^2$ and from this and Eqn. [A.2.13] it may be shown that

$$\mu_a(r) = \frac{G(1/r_1^2 - 1/r_2^2}{4\pi\Omega} \frac{n[1 - (r_1/r_2)^{2/n}](r/r_1)^{2/n-2}}{1 - (r_1/r_2)^2} \quad \text{[A.2.14]}$$

REFERENCES

1. Bingham, E. C., *Fluidity and Plasticity*, McGraw-Hill, New York, 1922.
2. Ostwald, W., *Kollaidzschr.* 1926 **38** 261.
3. Reiner, M., *Deformation and Flow*, Lewis, London, 1949.

4. Pryce-Jones, J., *Coll. Zeits.* 1952 **96** 129.
5. Freundlich, H. and Julisberger, F., *Trans. Faraday Soc.* 1935 **31** 920, 24.
6. Maxwell, J. C., *Phil. Trans.* 1867 **49** 157.
7. Schofield, R. K. and Scott-Blair, G. W., *Proc. Roy. Soc.* 1932 **A138** 707.
8. Oldroyd, J. G., *Proc. Roy. Soc.* 1953 **A218** 122.
9. Fröhlich, H. and Sack, R., *Proc. Roy. Soc.* 1946 **A185** 415.
10. Toms, B. A. and Strawbridge, D. J., *Trans. Faraday Soc.* 1953 **49** 1225.
11. Alfrey, T., *Mechanical Properties of High Polymers*, Interscience, New York, 1948.
12. Wratten, R., *Proc. 2nd. Intern. Congr. Rheol.* 1953, p. 181 (Butterworth, London, 1954).
13. Krieger, I. M. and Maron, S. H., *J. Appl. Phys.* 1954 **25** 72.
14. Piper, G. H. and Scott, J. R., *J. Sci. Instrum.* 1945 **22** 206.
15. Mooney, M. and Ewart, R. H., *Physics* 1934 **5** 350.
16. Weltmann, R. N., N.A.C.A. Techn. Note 3510 (1955).
17. Oldroyd, J. G., *J. Colloid Sci.* 1949 **4** 333.
18. Mooney, M., *J. Rheology* 1931 **2** 210.
19. Metzner, A. B. and Reed, J. C., *A.I.Ch.E. Journal* 1955 **1** 434.
20. Ambrose, H. A. and Loomis, A. G., *Physics* 1933 **4** 265.
21. Green, H., *Industrial Rheology and Rheological Structures*, Chapman and Hall, London, 1949.
22. Green, H. and Weltmann, R. N., *Industr. Engng. Chem.* (*Anal*) 1943 **15** 201; 1946 **18** 167.
23. Leadermann, H. *Elastic and creep properties of filamentous materials and other high polymers*, Textile Foundation, Washington, 1943.
24. Ferry, J. D., *Rheology*, Vol. II, Chap. 11, Academic Press, New York, 1958.
25. Markovitz, H. *et al.*, *Rev. Sci. Instrum.* 1952 **23** 430.
26. Van. Wazer, J. R. and Goldberg, H., *J. Appl. Phys.* 1947 **18** 207.
27. Mason, W. P., *Trans Amer. Soc. Mech. Engrs.* 1947 **69** 359.
28. Andrews, R. D., *Industr. Engng. Chem.* (*Anal*) 1952 **44** 707.
29. Swartzl, F. & Staverman, A. J., *Physica,'s Grav.* 1952 **18** 791.
30. Roberts, J. E., *Proc. 2nd. Intern. Congr. Rheol., Oxford* 1953.
31. Weissenberg, K., *Nature, Lond.* 1947 **159** 310.
32. Caldwell, D. H., and Babbitt, H. E., *Trans. Amer. Inst. Chem. Engrs* 1941 **37** 237.
33. McMillen, E. L., *Chem. Engns Progr.* 1948 **44** 537.
34. Hedström, B. O. A., *Industr. Engng. Chem.* 1952 **44** 651.
35. Ooyama, Y. and Ito, S., *Chem. Eng.* (*Japan*) 1950 **14** 96.
36. Williamson, R. V., *Indstr. Engng. Chem.* 1929 **21** 1, 108.
37. Campbell, L. E., *J. Soc. Chem. Industr. Lond.* 1940 **59** 71.
38. Powell, R. E. and Eyring, H., *Nature, Lond.* 1944 **154** 427.
39. Christiansen, E. B., Ryan, N. W. and Stevens, W. E., A.I.Ch.E. Journal 1955 **1** 544.
40. Scott-Blair, G. W., *A Survey of General and Applied Rheology*, Pitman, London, 1949.
41. Alves, G. E., Boucher, D. F. and Pigford, R. L., *Chem. Engng Progr.* 1952 **48** 385.
42. Metzner, A. B., *Chem. Engng Progr.* 1954 **50** 27.
43. Weltmann, R. N., *Industr. Engng. Chem.* 1956 **48** 386.
44. Winding, C. C., Baumann, G. P. and Kranich, W. L., *Chem. Engng Progr.* 1947 **43** 527, 613.
45. Dodge, D. W. and Metzner, A. B., *A.I.Ch.E. Journal* (to be published). Also *Rheologica Acta*, Band 1 (August 1958) p. 205
46. Oldroyd, J. G., *Proc. 1st Inst. Congress on Rheology* Scheveningen 1948 p. 130.
47. Schiller, L., *Z. Angew. Math. Mech.* 1922 **2** 96.
48. Bogue, D. C., *Industr. Engng. Chem. Symposium*, Pittsburgh, Dec. 1958.
49. Toms, B. A., *Proc. 1st Intern. Congr. Rheol.*, Scheveningen, 1948.
50. Weltmann, R. N. and Keller, T. A., N.A.C.A. Techn. Note 3889 (1957).
51. Fredrickson, A. G. and Bird, R. B., *Industr. Engng. Chem.* 1958 **50** 347.
52. *Industr. Engng. Chem.* 1953 **45** 969.
53. Rabinowitsch, B., *Z. Phys. Chem.* 1929 **A145** 1.

54. Yoshida, T., Hayashida, K., Kobayshi, K. and Tanaka, H. *Chem. Eng.* (*Japan*) 1957 **21** 26
55. Mori, Y. and Matsumoto, T. K., *Proc. 3rd Intern. Congress Rheol*, 1958. p. 240.
56. Mori, Y. and Ototake, N., *Chem. Eng.* (*Japan*) 1953 **17** 224; 1954 **18** 221.
57. Nickolls, K. R. and Colwell, R. E., *Industr. Engng. Chem. Symposium*, Pittsburgh, Dec. 1958.
58. Gaskell, R. E., *J. Appl. Mech.* 1950 **17** 334.
59. Kay, J. M., *Introduction to Fluid Mechanics and Heat Transfer* Camb. Univ. Press, 1957.
60. Leveque, M. A., *Ann. Min.* 1928 **13** 201.
61. Lyche, B. C. and Bird, R. B., *Chem. Eng. Sci.* **1956 6** 35.
62. Pigford, R. L., *Chem. Engng. Progr. Symposium Service No.* **17**. 1955 **51** 79.
63. Metzner, A. B., Vaughn, R. D. and Houghton, G. L., *A.I.Ch.E. Journal*. 1957 **3** 92
64. Gee, R. E. and Lyon, J. B., *Industr. Engng. Chem.* 1957 **49** 956.
65. Winding, C. C., Dittmann, F. W. and Kranich, W. L., Cornell Univ. Report, Ithaca, 1944.
66. Orr, C. and Dalla Valle, J. M., *Chem. Engng. Progr. Symposium* Service No. 9. 1954 **50** 29.
67. Metzner, A. B. and Friend, W. L., *A.I.Ch.E. Journal* 1958 **4** 393
68. Metzner, A. B. and Friend, P. S. *Industr. Engng. Chem. Symposium*, Pittsburgh, Dec. 1958.
69. Rushton, J. H., *Chem. Engng Progr.* 1954 **50** 587.
70. Hixon, A. W. Drew, T. B. and Knox, K. L., *Chem. Engng Progr.* 1954 **50** 592.
71. Metzner, A. B. and Otto, R. E., *A.I.Ch.E. Journal* 1957 **3** 3.
72. Lee, R. E., Finch, C. R. and Wooledge, J. D., *Industr. Engng. Chem.* 1957 **49** 1849.
73. Foresti, R. J., *Industr. Engng. Chem. Symposium*, Pittsburgh, Dec. 1958.
74. Brown, G. A. and Petsiavas, D. N., Paper presented at the New York A.I.Ch.E.
75. Meeting, December 1954.
76. Schultz-Grunow, F., *Chem.-Ing.-Tech.* 1954 **26** 18.
77. Calderbank, P. H. and Moo-Young, M. B., *Trans. Instn Chem. Engrs, Lond.* 1959 **37** 26.
78. Weissenberg, K., *Proc. 1st Intern. Cong. Rheol.*, Scheveningen, 1948.
79. Russell, R. J., Ph.D. Thesis, London University, 1946.
80. Roberts, J. E. and Jobling, A. To be published.
81. Oldroyd, J. G., *J. Colloid Sci.* 1949 **4** 333.
Toms, B. A., *J. Colloid Sci.* 1949 **4** 511.

INDEX

Hgn SP
8304

NOTTINGHAM
UNIVERSITY
AGRICULTURAL
SCIENCE
LIBRARY